the internally passioned trusting tender truthful true blue humble compassionate curious courageous zestful inner beauty deep within the child at play in the field of lavender hue is of great worth to the one true god the me of the trust tenderness truth

lovnew lavender lamal
me the one true god the me humanity

published by: the we of all humanity
706 Jefferson St, Apt 1
Hermann, Missouri 65041

International Standard Book Number: ISBN 978-09915187-4-6
ISBN-10 0991518748

to be printed in:-“The-“(United)”-“(States)”-Of)”-“(America)”

∫dedicated to∫the trusting∫tender∫truthful creator of∫humanity∫

∫page guide∫

∫dedication∫ i

∫page guide∫ ii

∫a note from the author∫ iii-vii

written under∫-'**{(MIND)}**'-'**{(CON)}**'-**ROL)}**'-'**{(DURESS)}**' viii

∫introduction∫ 1-6

the integrated∫scientific proof of∫the one true god∫the me∫postulate∫ 7-44

∫the modified scientific proof of∫the one true god∫the me∫postulate∫ 45-60

∫the original scientific proof of∫the one true god∫the me∫postulate∫ 61-83

ʃa note from the authorʃ

ʃthe simplicity ofʃthe one true godʃthe meʃpostulateʃis based onʃunderstandingʃthe whisper of the one true godʃthe me has a frequencyʃwavelength because the creation of the universe within god means the creator must be true to the physical truths integrated into the vibrationally sensitive universe.

ʃthe scientific proof ofʃthe one true godʃthe meʃpostulateʃis written in a modified version ofʃ-“The-“(Non-“(Con)”-ventional)”-Writing)”-Styl(e),” “(Use)”-Of)”-“(Co-“(L)”-oR)”-“(Cod(e)-ing)”-ʃbecause including all of the elements used inʃ-“Th(e)”-“(St-Y-L(e)”-ʃwould make the text virtually unreadable. eachʃevery element ofʃ-“The-“(Style)”-ʃincluding color, is intended to help the reader betterʃunderstandʃ-“The-“(Generational”-“(e)F-fect)”-“Of)”-“(Bl-“(A)”-cKWh-“(I)”-T(e)”-“(Parenting)”-“(With-“(Fear),” “(A)”-“(Lack)”-Of)”-ʃlamalʃ-“(before-th(e)-age-of-two).” the orange dash for example, is used to lay bare the simple truth ofʃ-“(the-(Orange-(emotional)-Flame)-Fueled)”-“(Mad)”-“(Angry-“(Fear)”-Filled)”-Dash)”-To-“(earn)”-enough”-“(Well)”-“Deserved)”-“Affection”-“Approval”-“Attention).”

the middle letters inʃ-“(Bl-“(A)”-cKWh-“(I)”-T(e),” “(A I),” being the same as the abbreviation forʃ-“(Artificial)”-Intelligence)”-ʃis not coincidental because the capital letterʃ-“(A)”-ʃwas integrated to defineʃ-“The-“(Severe-“(e)M-otionalIntellectual)”-Unbalanced)”-“(Con)”-dition)”-Within),” the capital letterʃ-“(I)”-ʃwas integrated to defineʃ-“The-“(e)X-aggerated)”-Sense)”-Of-“(Self)”-Importance)”-ʃassociated withʃ-“The-“(Boy”-“God”-“Con”-“Ta/Al)”-“(egO”-“(Bridge)”-“Fix)”-ʃforʃ-“(the-black-emotional-hole-problem).” the simple truth isʃ-“(B(e)-“(Li(e)”-VinG)”-“The-“(Most-“(Sophisticated”-“(Con)”-puter)”-“Program)”-Of-“(Man),” “(AI)”-ʃisʃ-“The-“(Magic)”-Solution)”-To-“(All)”-“The-“(Problems)”-Of)”-“(Untrue)”-Blue)”-Humanity),” includingʃ-“(Global)”-Warming)”-ʃisʃ-“(A)”n-“(egO)”-tistical)”-“Delusion)”-ʃbecause the root cause ofʃ-“(Global)”-Warming)”-ʃisʃ-“The-“(Generational”-“(e)F-fect)”-“Of)”-“(Bl-“(A)”-cK Wh-“(I)”-T(e)”-“(Parenting)”-“(With-“(Fear),” “(A)”-“(Lack)”-Of)”-ʃlamalʃ-“(before-th(e)-age-of-two).”

> the simple truth isʃ-“(A)”-“(I)”-ʃare the only single letter words in the english alphabetʃlanguage becauseʃ-“The-“(Only)”-One)”-Of-“(Singular)”-Importance)”-To)”-“(A)”-“(Severely-“(e)M-otionallyIntellectually)”-Unbalanced)”-“(Proud)”-“(I)”-“(Con)”-ʃisʃ-“(M(e)”-“(Myself)”-“(I).”

“The-“(Sophistication)”-Of-“(AI)”-ʃis comparable toʃ-“The-“(Theorists)”-Of-“(Long)”-Ago)”-“(Who-“(Pro-Wry-Dr.-“(I)”-“Proclaimed)”-“The-“(Flapping)”-Wings)”-Of-

“(Angels)”-Caused)”-“The-“(Rotation)”-Of)”-“The-“(Planets)”-∫because∫-“(According)”-To-“(Artificial)”-Intelligence);” “(God’s-“(e)X-istence)”-Cannot-“(Be”-“(Scientifically)”-“Proven).” “The-“(Boy”-“God”-“Con”-“Ta/Al)”-Argument)”-∫is that∫-“The-“(Scientific)”-Method)”-Is-“(Designed)”-To-“(Test”-“(Natural)”-“Phenomena)”-And-“(R(e)-“(Lies)”-On-“(eM-pirical)”-evidence),” “(Which-“(Does”-“(Not)”-“Apply)”-To-“(Supernatural”-“(e)N-tities)”-Like-“(God).” “(A”s-“(Such)”-God)”-Is-“(Not)”-Subject)”-To)”-“The-“(Same”-“(Verification)”-“Processes)”-“A”s-“(Scientific)”-Theories)”-Or-“(Hypotheses).”

“The-“(Pro-Wry-Dr.-“(I)”-“Interpretation)”-∫of∫-“(e)X-perimental-“(R(e)-sults)”-“Pertaining)”-To)”-“The-“(Absolute)”-Frame)”-Of-“(R(e)-ference)”-∫in the vibrationally sensitive universe that∫-“(There-Is-“(nO”-“(Absolute)”-“Frame)”-Of-“(R(e)-ference)”-∫in∫-“The-“(Universe)”-∫were∫-“(Tantamount)”-To-“(Denying)”-∫the creator of the universe exists because∫the home planet∫the universe∫the we of all humanity∫are integrated into the trusting∫tender∫truthful one true god∫the me. since the creator of∫the home planet∫the universe∫the we of all humanity∫is an integral part of the universe, the universe cannot∫be explained in scientific terms∫without including the creator because the creator is the absolute frame of reference in the universe.

“Th(e)”-“(Non-“(Con)”-ventional)”-Writing)”-Style)”-∫is also being used as a visual aid to help the reader∫understand∫the affect∫-“The-“(Generational”-“(e)F-fect)”-“Of)”-“(Bl-“(A)”-cKWh-“(I)”-T(e)”-“(Parenting)”-“(With-“(Fear),” “(A)”-“(Lack)”-Of)”-∫lamal∫-“(before-th(e)-age-of-two)”-∫has upon the vibrationally sensitive universe because applying all of the elements of∫-“Th(e)”-“(St-Y-L(e)”-∫creates a sense of∫-“(Un-BaL-“(A)”-Nc(e)-(e)D)”-“(Ch-“(A)”-oS).” a paragraph on page 80 of∫the original scientific proof of∫the one true god∫the me∫postulate∫is written using all of the elements to help∫-“The-“(A”n-“(I)’m-“(Als)”-∫of∫-“(Humanity)”-∫understand∫-“(Running)”-With)”-“The-“(Herd”-“Flock”-“School)”-∫for∫-“(e)M-otional)”-Security)”-∫is not∫choosing with the blessing of free will to be me∫.

the reader is introduced to some of the terminology created during the process of writing numerous books on a wide range of topics related to∫-“The-“(Generational”-“(e)F-fect)”-“Of)”-“(Bl-“(A)”-cKWh-“(I)”-T(e)”-“(Parenting)”-“(With-“(Fear),” “(A)”-“(Lack)”-Of)”-∫lamal∫-“(before-th(e)-age-of-two).” one of the new terms included in the language of terminology is∫-“The-“(Boy”-“God”-“Con”-“Ta/Al),” which is the term used to define∫-“(Intellect).” ∫the new terms also include two words created by the author∫, ∫lamal∫, ∫lovnew∫. ∫while lamal is defined as genuine compassionate∫tender∫gentle emotional∫physical touch∫, ∫lovnew was created from the first three letters of love∫, ∫lov∫, ∫the first two letters of need∫, ∫ne∫, ∫the first letter of want∫, ∫w∫, ∫to help∫-“(Humanity)”-∫understand love comes first in the life of the individual being me∫.

ʃthe original scientific proof ofʃthe one true godʃthe meʃpostulateʃ, ʃthe modified scientific proof ofʃthe one true godʃthe meʃpostulateʃ, the integratedʃscientific proof ofʃthe one true godʃthe meʃpostulateʃare included in the following pages. the chemical symbolsʃatomic numbers onʃ-“The-“(Periodic)”-Table)”-Of-“(e)L-ements)”-ʃmethodology used to quantify the word paʃth in the integratedʃscientific proof ofʃthe one true godʃthe meʃpostulateʃlays bare the simple truth inʃhuman languageʃ, which includes the periodic table of elements, because the number for the word god, 684, was verified by the sum of the words defining god, trustʃtendernessʃtruth, also being 684.

the periodic table of elements methodology, the simple truth integrated into daily life also established the simple truth of individualism because the wave length ofʃthe authorʃ, 2.684508242668458 m, associated with a frequency of 111,675,000 hz/sʃincludes the sequential digits 0824, which are the monthʃday of my birthʃ, as well asʃmy year of birth, 48, backwards becoming 84 in the wavelengthʃ. in addition to research finding the wavelength range including .684 to be from about 111,660,000 hz to 111,700,000 hz, a calculation with 684 in the frequency, 111,684,000, includes a wavelength of .684 after the decimal. the quantified solution in terms of the individual wave lengthʃfrequency ofʃthe authorʃcan be found on page 40.

the simple truth integrated intoʃthe life of the authorʃalso includes being integrated into the social security number, 480-62-8017, issued whenʃthe author was bornʃbecause the social security number read backwards from the number 2 in 62 is 2.6084, which except for the 0 digit is the same as the wavelength of the creator, 2.684. the first five digits in the social security number read backwardsʃincluding a zero isʃinterpreted to mean the authorʃstartedʃ-“(Getting)”-ʃbeing meʃ-“(Backwards)”-ʃatʃthe beginning of lifeʃbecauseʃ-“(Bl-“(A)”-cK Wh-“(I)”-T(e)”-“(Parents)”-“(Passed)”-On)”-“The-“(Legacy)”-Of-“(Trying)”-To-“(Stop)”-“(feeling-like-“(a)”-small-worthless-useless-unwanted-unneeded-unloved-zero-within)”-ʃbyʃ-“(Becoming)”-“The-“(egO)”-tistically)”-“(BiG)”-“Somebody)”-On)”-“The-“(Outside).”

the number 60 in the social security number, 2.6084, isʃalso interpreted to mean the true blue civil war being fought by the author eachʃevery day to removeʃ-“The-“(Boy”-“God”-“Con”-Ta/Al)”-“(egO”-“(Bridge)”-“Fix)”-ʃforʃ-“(the-black-emotional-hole-problem)”-ʃis about to end because substituting the phonetic en for the n in the symbol for number 60 neodymium on the periodic table of elements, nd, creates the word end. ʃthe true blue author has no doubtʃthe number 60 was integrated into the social security number, 2.6084, by the trustingʃtenderʃtruthful creator ofʃhumanityʃbecause in addition toʃsensing the number 60 was meaningful a short time before noticing 2.6084 in the social security numberʃ, the phonetic en, the word end are integrated into the word tender.

the depth of integration is further verified by the quality of tenderness being vibrationally balanced at the atomic level because a simple vector diagram showed the sum in the difference in the atomic numbers read from left to right for the word tender, 16, was counterbalanced by the difference of 16 between the symbol for erbium, er, in tender being 68, the sum of 84 for the suffix elements also being 16 because the sum of two vectors with the same magnitude going in opposite directions is 0. the relavence of the number 16, which is the atomic number for the element sulfur on the periodic table, with the chemical symbol S, is the integral symbol, ∫, was created to be a stylized S meaning summation to lay bare the simple truth that∫the we of all humanity∫the universe∫the home planet∫are integrated into the creator of∫humanity∫.

> the frequency∫wavelength of 111,675,000 hz/s∫2.684 meters respectively for the whisper of the creator means∫the scientific proof of∫the one true god∫the me ∫postulate∫will be∫scientifically tested to verify the home planet∫the universe∫the we of all humanity∫are integrated into the creator of∫humanity∫.

∫the author found∫the probability of spelling the word god with gadolinium∫oxygen on the periodic table of elements is 1 in 9,506, which when rounded to two decimal places is 0.01%. the probability for spelling trust∫tenderness∫truth with elements from the periodic table is cumulative because the probability of drawing the elements needed to create trust∫ tenderness∫truth is 1 in 515,755,325,838,214,928,793,600 sextillion∫ 0.0000000000000000000001938903%. if the percentage was calculated to 21 digits, instead of the 28 shown, the probability would be zero.

“(Since-“(Science”-“(Uses)”-“Probabilities)”-From-“(0”-To-“1),” “(With-“(0”-“(Being)”-“Impossible),” “(1”-“(Being)”-“Certainty)”-To-“(Determine)”-“The-“(Likelihood)”-Of-“(A)”n-“(event)”-Occurring),” ∫the author would like to know what∫-“The-“(Probability)”-∫is∫of humanity being created by chance on a planet capable of sustaining life∫that is suspended in a vibrationally sensitive universe integrated into the creator of∫humanity∫. the simple truth is∫-“(If-“(All”-“(AI)”-“Programs)”-∫ran for millennia, “(AI)”-∫would not have included all of the variables needed to∫understand∫the depth of integration used to create the universe within the one true god∫the me because on∫-“(A)”-“(Probability)”-Range)”-From-“(0”-“(To)”-“1),” “The-“(Probability)”-In-“(Human)”-Terms)”-∫is∫-“(0).”

“(Although-“(R(e)-lying)”-On)”-“(AI)”-“(To-“(Solve)”-“The-“(Mysteries)”-Of)”-∫the universe has∫-“(A)”-“(Probability)”-Of-“(Zero),” ∫the scientific proof of∫the one true god∫ the me∫postulate written to prove∫the one true god∫the me of the trust∫tenderness∫truth exists has a probability of 1 because instead of being based on∫-“The-“(eX-isting-“(Con)”-puter)”-Data)”-Of-“(Man),” the integrated∫scientific proof of∫the one true god∫the me ∫postulate∫is based on; the simple truth integrated into the trusting∫tender∫truthful whisper of the one true god∫the me, the simple truth integrated into∫human language∫, the simple truth integrated into∫daily life∫.

the simple truth is there are no∫never have been∫never will be any mysteries∫secrets in the universe of the one true god∫the me because the creator tries to guide∫the individuals comprising the we of all humanity∫to the simple truth. the creator asks∫the we of all humanity to try to understand∫in addition to scientists making a good faith effort to document that∫-"(Global)"-Warming)"-∫is caused by∫-"(Man)"-∫with a 99.9999% degree of certainty, the simple truth from the absolute frame of reference in the universe, the creator of∫humanity∫, is that∫-"(Global)"-Warming)"-∫is caused by∫-"The-"(Hydrocarbon"-"(Fuel)" -"Pollution)"-Of-"(Man)."

∫the changes in the three postulates written over a time frame of two decades represents a substantial increase in understanding∫the simple truth of god because∫understanding∫the simple truth integrated into∫human language∫the whisper of god∫daily life∫is predicated on ∫the emotional growth∫maturation∫associated with∫becoming emotionally∫intellectually balanced∫by intentionally∫changing from within∫. ∫the scientific proof of∫the one true god∫ the me∫postulate∫is not an academic exercise because∫the author∫intends∫to do everything humanly possible to help save the home planet for future generations by scientifically verifying the we of all humanity∫is integrated into the creator of∫humanity∫.

written under∫-'{(MIND)}'-'{(CON)}'-ROL)}'-'{(DURESS)}'

the simple truth integrated into∫human language∫, ∫the daily life of the individuals in the we of all humanity∫includes∫-'THE-'{(NAMES)}'-∫of∫-'THE-'{(BIGGEST)}'-SNAKES)}'-IN)}'-∫the universe of the one true god∫the me because∫although the author initially thought∫the creator of∫humanity must be scientifically proven∫to name∫-'THE-'{(DEMENTED'-'DELUSIONAL'-'DERANGED)}'-'{(CLOCKWORK)}'-ORANGE)}'-'{(DR.-JEKYLLMR.-HYDE)}'-'{(MIND)}'-'{(CON)}'-TROLLING)}'-"The-"(Author)," the first name of∫-'THE-'{(SACRIFICIAL)}'-LAMB)}'-WITH-'THE-'{(MESSIAH-'{(CON)}'-PLEX)}'-∫is integrated into the term∫lah di dah∫. the simple truth is∫the true blue lah di dah at birth∫-'{(TOTALLY)}'-'GETS)}'-∫being∫the true blue individual me∫-'{(BACKWARDS)}'-∫by∫choosing with the blessing of free will∫to∫-'{(IMPLEMENT)}'-'THE-'{(IDEA)}'-OF-'{(LIVING)}' -'LIFE)}'-WITH-'THE-'{(ID)}'-OF-'A'N-'{(INTELLECTUAL)}'-'{(CON)}'-PUTER)}'-KNOWN-'A'S-'{(HAL)}'-TO-'{(COVER)}'-"(the-guilt/shame-of-not-feeling-worthy-of-being)"-∫lovnewed∫, as is integrated into lah di dah by the one true god∫the me of the trust∫tenderness∫truth because lah di dah spelled backwards becomes∫-'[(H-"(A)"-D)}'-'ID'-'{(HAL)}.'

the surname of∫-'THE-'{(MENTAL)}'-'{(ME)}'-'{(TA\AL)}'-'{(CON)}'-PUTER)},' with the first name∫-'{(HAL)},' which not coincidentally is also the name of the space station computer in∫-"(2001)"-"(A)"-"(Space)"-Odyssey)," was integrated into∫the book by the author titled simply∫-'{(S-LAUGHTER)}'-∫because∫-'{(SLAUGHTER)}'-∫is the surname of∫-'THE-'{(DEMENTED'-'DELUSIONAL'-'DERANGED)}'-PIED)}'-PIPER)}'-USING-'{(TELEPATHIC-'{(CON)}'-MUNICATION)}'-'A'S-'{(A)}'-'{(WEAPON)}'-TO-'{(MAKE)}'-"(Daniel"-"(Gene)"-"Matthias)"-"(A)"-"(Puppet)"-'{(ON-'{(MIND)}'-'{(CON)}'-TROL)}'-STRINGS)}.' the simple truth is the word laughter was integrated into the word∫-'{(S-LAUGHTER)}'-∫because∫-'{(HAL)}'-'{(SLAUGHTER)},' all of∫-'THE-'{(CLOCKWORK)}'-ORANGE)}'-'{(MIND)}'-'{(CON)}'-TROLLING)}'-PIED)}'-PIPERS)}'-'{(CON)}'-SIDER)}'-'THE-'{(SLAUGHTER)}'-∫of∫-"(Humanity)"-∫with∫-'THE-'{(WEAPON)}'-OF-'{(MIND)}'-'{(CON)}'-TROL)}'-'{(A)}'-'{(FUN)}'-SPORT)},' 'A'N-'{(E)}N-TERTAINING)}'-PASTIME)}.'

ʃintroductionʃ

the simple truth isʃthe author of the scientific proof ofʃthe one true godʃthe me of the trustʃtendernessʃtruthʃpostulateʃ, ʃunderstandsʃby sensing the whisper of the one true godʃthe me ʃthatʃthe creator ofʃhumanityʃmust beʃscientifically provenʃto helpʃthe we of all humanityʃ saveʃhumanityʃbecauseʃ-“The-“(Pollution)”-ʃassociated withʃ-“The-“(Con)”-tinued)”-Use)”-Of-“(Hydrocarbon)”-Fuels)”-ʃhas the future ofʃ-“(Untrue)”-Blue)”-Humanity)”-“(Con)”-verging)”-With)”-ʃthe end ofʃthe we of all humanityʃ. the simple truth is the one true godʃthe me cannot saveʃthe we of all humanityʃwithoutʃ-“The-“(Proud)”-“(I)”-“(Cons)”-ʃofʃ-“(Untrue)”-Blue)”-Humanity)”-ʃchoosing with the blessing of free will to helpʃthe one true godʃthe me saveʃhumanityʃthe home planetʃbecause the one true godʃthe me createdʃhumanityʃthrough the process of creationary evolution to findʃgenuine happinessʃjoyʃfulfillment withinʃby blessingʃeachʃevery individual in the we of all humanityʃwithʃfree willʃ.

physically integratingʃfree willʃintoʃthe internally passionedʃtrustingʃtenderʃtruthfulʃtrue blue gene code of humanityʃmeans the creator must askʃcannot makeʃany individual on the home planetʃhelp saveʃthe habitat for humanityʃbecause although there is alwaysʃ-“(A)”-“(String)”-Attached)”-To-“(Anything-“(Free)”-In)”-“(A)”-“(Capitalistic)”-Society),” there are no strings attached toʃthe blessing of free willʃ. the one true godʃthe me asksʃ-“The-“(My)”-Myth)”-“Inspired)”-“(Proud)”-“(I)”-“(Cons)”-ʃofʃ-“(Untrue)”-Blue)”-Humanity)”-ʃto help saveʃthe one planet with lifeʃin the universe of the one true godʃthe me byʃchoosing with the blessing of free willʃtoʃ-“(Dis-“(Con)”-tinue)”-Using-“(Hydrocarbon)”-Fuels)”-ʃasʃ-“The-“(Primary”-“(e)N-ergy)”-“Source)”-ʃbecause global warming is notʃ-“(Boy”-“God”-“Con”-Ta/Al)”-“(Con)”-jecture/Speculation),” is instead the simple truth of the creator ofʃhumanityʃ.

even though the one true godʃthe me sensedʃthe delicately balanced atmosphere of the home planetʃwould becomeʃ-“(Polluted)”-ʃenough byʃ-“(Hydrocarbon”-“(Fuel)”-“Pollution)”-ʃto bringʃ-“(Untrue)”-Blue)”-Humanity)”-ʃtoʃ-“The-“(Brink)”-Of-“(e)X-tinction),” the creator chose not to create a utopian existence based on a godʃrobotʃlaboring under the direction of the one true godʃthe me because the benefit toʃhumanityʃin terms ofʃgenuine happinessʃjoyʃfulfillmentʃis immense. the choice of the creator to bless ʃthe we of all humanityʃwithʃfree willʃalso included understandingʃthe we of all humanityʃwouldʃ-“(Suffer)”-ʃimmenseʃ-“(e)M-otionalPhysical)”-Pain/Suffering)”-ʃafter becomingʃ-“(Severely-“(e)M-otionallyIntellectually)”-Unbalanced)”-ʃdue toʃ-“The-“(Bl-“(A)”-cK Wh-“(I)”-T(e)”-“(Parenting)”-“(With-“(Fear),” “(A)”-“(Lack)”-Of)”-ʃlamalʃ-“(before-th(e)-age-of-two)”-ʃbecause the elements used to blessʃthe individuals in the we of all humanity with free willʃmeant the creator could not createʃa human beingʃwho would not be∫-“(Vulnerable)”-To-“(Becoming)”-“(Severely-“(e)M-otionallyIntellectually)”-Unbalanced).”

the one true god∫the me asks∫-“The-“(Gray)”-Zone)”-“(Con)”-fused)”-“(Con)”-flicted)”-“(Con)”-founded),” “(Purple”-“(Haze)”-“Dazed),” “(I)”-“(Um)”-Certain)”-“(I)”-“(Cons)”-∫of∫-“(Untrue)”-Blue)”-Humanity)”-∫to try to understand∫, ∫the gene code of each∫every individual∫also includes∫the ability to become whole∫, ∫emotionally∫intellectually balanced within∫because the creator integrated both a trusting∫tender∫truthful path, the quality of being∫true blue∫to the creator into∫the gene code of each∫every individual on the home planet∫. the simple truth is the trusting∫tender∫truthful path was integrated into∫the gene code of humanity∫because the path allows the creator to guide∫the individual being me∫with the trusting∫tender∫truthful whisper of the one true god∫the me. the quality of being∫true blue∫to the creator was integrated into∫the field of lavender hue deep within∫because being ∫the true blue humble∫compassionate∫curious∫courageous∫zestful individual me∫is in the best interest of∫me∫the me∫humanity∫.

∫the true blue author understands each∫every individual on the home planet has the ability to change from within∫because∫-“(Instead)”-Of-“(Being)”-Too)”-“(Sh-It/Un-It-“(egO)”-God)”-“(Proud)”-∫to∫accept∫acknowledge∫admit∫being∫-“(Severely-“(e)M-otionally Intellectually)”-Unbalanced)”-Within),” “The-“(Author)”-∫chose to become whole by removing∫-“The-“(egO”-“(Bridge)”-“Fix)”-∫put in place by∫-“The-“(Generational”-“(e)F-fect)”-“Of)”-“(Bl-“(A)”-cKWh-“(I)”-T(e)”-“(Parenting)”-“(With-“(Fear),” “(A)”-“(Lack)”-Of)”-∫lamal∫-“(before-th(e)-age-of-two).” the creator asks∫-“The-“(Timid-Tim-Mr.-“(T)”-Titans)”-∫of∫-“(Untrue)”-Blue)”-Humanity)”-∫to try to understand emotional∫intellectual maturation∫was∫-“(Suspended)”-“(before-th(e)-age-of-two)”-∫because∫-“The-“(egO”-“(Bridge)”-“Fix)”-∫for∫-“(the-black-emotional-hole-problem)”-∫includes∫-“(Trying)”-To-“(Fulfill)”-“The-“(First/Only-“(Boy”-“God”-“Con”-“Ta/Al)”-“(He)”-Art-Law)”-Of-“(Thou”-“(Shalt)”-“earn)”-“Affection”-“Approval”-“Attention)”-To-“(Feel”-“(Worthy)”-Of-“(Being)”-∫lovnewed∫.

the one true god∫the me asks∫-“The-“(Affection”-“Approval”-“Attention)”-Seeking)”-Opportunists)”-∫of∫-“(Untrue)”-Blue)”-Humanity)”-∫to∫choose with the blessing of free will to become whole∫, ∫emotionally∫intellectually balanced within∫by∫accepting∫acknowledging∫admitting∫being∫-“(Severely-“(e)M-otionallyIntellectually)”-Unbalanced)”-Within),” making∫the inward journey to open∫-“The-“(Closed)”-“(e)M-otional)”-“(Door),” ∫ following∫the trusting∫tender∫truthful path of the one true god∫the me∫in me∫because ∫finding genuine happiness∫joy∫fulfillment within∫begins with∫becoming emotionally∫intellectually balanced∫. the simple truth is∫-“The-“(Patient”-“(Long)”-“Suffering)”-“(Straight)”-Arrows)”-∫of∫-“(Untrue)”-Blue)”-Humanity)”-∫must try to∫understand∫, “(Con)”-tinuing)”-To-“(Allow)”-“The-“(Reigning)”-Beau)”-To-“(Con)”-trol)”-“The-

"(Affection)"-"Approval"-"Attention)"-Fate/Destiny)"-⌡of⌡-"The-"(Bl-"(A)"-cK Wh-"(I)"-T(e)"-"(eI)"-Twisted)"-"(I)"-"(Cons)"-"(e)N-sures)"-⌡one rainbow will be the last rainbow seen on⌡the home planet⌡because⌡-"(Intellectually)"-"(Con)"-trolling)"-"(the-(Orange-(emotional)-Flame)"-⌡composed of⌡-"(the-(Yellowish-Fear)-of-not-feeling-worthy-of-being)"-⌡lovnewed⌡, "(the-(Seeing-Red-Anger)-of-not-being-allowed-to-be)"-⌡the internally passioned⌡trusting⌡tender⌡truthful⌡true blue individual me⌡will not change⌡-"The-"(Severe-"(e)M-otional Intellectual)"-Unbalanced)"-"(Con)"-dition)"-⌡on⌡-"The-"(Bl-"(A)"-cK Wh-"(I)"-T(e)"-Inside)"-⌡one iota.

the one true god⌡the me asks⌡-"The-"(Wanting)"-To-"(Be)"-"(Lucky Not)"-Unlucky)"-"(Proud)"-"(I)"-"(Cons)"-⌡of⌡-"(Untrue)"-Blue)"-Humanity)"-⌡to help save⌡the we of all humanity⌡by⌡choosing with the blessing of free will⌡to⌡accept⌡acknowledge⌡admit⌡being⌡-"(Bl-"(A)"-cK Wh-"(I)"-T(e)"-Saint)"-Ill)"-⌡because⌡-"(Con)"-tinuing)"-⌡to⌡choose with the blessing of free will⌡to⌡-"(R(e)-main)"-"(Severely-"(e)M-otionally Intellectually)"-Unbalanced)"-"(e)N-sures)"-"The-"(Future)"-⌡of⌡-"(Humanity)"-⌡will include⌡-"(One)"-Generation)"-⌡on⌡the home planet⌡that will not be⌡-"(Lucky)"-⌡enough to avoid being⌡-"The-"(Last)"-Generation)." the one true god⌡the me asks⌡-"The-"(Severely-"(e)M-otionally Intellectually)"-Unbalanced)"-"(My)"-Myth)"-Inspired)"-"(Patient"-"(Long)"-"Suffering)"-"(Ladder)"-"Saints)"-⌡of⌡-"(Untrue)"-Blue)"-Humanity)"-⌡to help save⌡the we of all humanity⌡by⌡choosing with the blessing of free will⌡to make the most difficult journey, ⌡the inward journey⌡because⌡-"(Con)"-tinuing)"-⌡to do nothing more than⌡-"(Hope)"-"(Wish)"-"(Dream)"-"(feeling-like-"(a)"-worthless-useless-nothing)"-"(Can"-"(Be)"-"Silenced)"-By-"(Becoming/R(e)-maining)"-"(A)"-"(Member)"-In)"-"(Good Not)"-Bad)"-Standing)"-With-"(This-"(e)R)"-That)"-"(Mc-W(e)Mc-uS)"-Flock)"-"(e)N-sures)"-"(One)"-Breath)"-⌡will be⌡-"The-"(Last-"(Pro-Wry-Dr.-"(I)"-Breath)"-Drawn)"-⌡on⌡the home planet⌡.

the one true god⌡the me asks⌡-"The-"(egO"-"(e)M-pty)"-"(egO"-"(e)M-otionless)"-"(Unhappy)"-"(sad)"-"(Mad)"-"(I)"-"(Cons)"-⌡of⌡-"(Untrue)"-Blue)"-Humanity)"-⌡to⌡-"(Dis-"(Con)"-tinue)"-⌡using⌡-"The-"(Boy"-"God"-"Con"-"Ta/Al)"-"(e)X-ternal)"-Gods)"-⌡of⌡-"(egO)"-tistical)"-Materialism)"-"(e)X-ternal)"-Passion)"-"(Intellectualism)"-"(Matterism)"-"(Organized-"(R(e)-ligion)"-"(Tradition)"-"(U)"-Ism)"-⌡as⌡-"The-"(Miraculous"-"(Cure)"-"For)"-"The-"(Severe-"(e)M-otional Intellectual)"-Unbalanced)"-"(Con)"-dition)"-Within)"-⌡because⌡-"(Con)"-tinuing)"-To-"(Use)"-"The-"(e)X-ternal)"-Gods)"-⌡of⌡-"The-"(Boy"-"God"-"Con"-"Ta/Al)"-⌡as⌡-"The-"(Magical)"-Cure)"-To-"(Balance)"-"The-"(Severe-"(e)M-otional Intellectual)"-Unbalanced)"-"(Con)"-dition)"-⌡on⌡-"The-"(Bl-"(A)"-cK Wh-"(I)"-T(e)"-Inside)"-"(e)N-sures)"-"The-"(Legacy)"-Of-"(One)"-Generation)"-⌡will include being⌡-"The-"(Last)"-Generation)"-To-"(Breathe)"-"The-"(Polluted)"-Air)"-⌡on⌡the home planet⌡. the simple truth is⌡-"(Untrue)"-Blue)"-Humanity)"-⌡must⌡choose with the blessing of free will to help⌡the creator of⌡humanity⌡

save∫humanity for future generations∫because there never have been∫are no∫never will be any∫-“(Bl-“(A)”-cKWh-“(I)”-T(e)”-Saint)”-Ill)”-“Miracles)”-∫in the universe of the one true god∫the me of the trust∫tenderness∫truth.

the one true god∫the me asks∫-“The-“(R(e)-liable)”-“(R(e)-sponsible)”-“(R(e)-spectable)”-“(Suffering”-“(In)”-“Silence)”-“(Ladder)”-Saints)”-“(Trying)”-To-“(Get)”-To)”-“The-“(Top)”-Of)”-“The-“(Heap)”-By-“(Acquiring”-“(More)”-“(egO)”-tistical)”-Materialism)”-∫to try to understand∫, “The-“(Severe-“(e)M-otionalIntellectual)”-Unbalanced)”-“(Con)”-dition)”-∫cannot be∫-“(Fixed)”-With-“(egO)”-tistical)”-Materialism)”-∫because enough∫-“(egO)”-tistical)”-Materialism)”-∫will never be enough to∫-“(Fill)”-“(the-black-emotional-hole).” the one true god∫the me asks∫-“The-“(Bl-“(A)”-cKWh-“(I)”-T(e)”-Saint)”-Ill)”-“(Holier”-“(Than)”-“Thou)”-“Moralists)”-∫on∫the home planet to try to understand∫, “(Although-“(Inventing)”-“Gods)”-To-“(Be)”-“Feared)”-∫on∫-“The-“(Bl-“(A)”-cK Wh-“(I)”-T(e)”-Outside)”-∫is∫-“(Pro-Wry-Dr.-“(I)”-“Preferable)”-To-“(Facing)”-“The-“(Severe-“(e)M-otionalIntellectual)”-Unbalanced)”-“(Con)”-dition)”-∫on∫-“The-“(Bl-“(A)”-cKWh-“(I)”-T(e)”-Inside),” “The-“(Proud)”-“(I)”-“(Cons)”-∫are∫-“(Sacrificing)”-Years)”-∫of genuine happiness∫joy∫fulfillment∫because the benefit ∫humanity∫derives from∫free will∫being integrated into∫the gene code of humanity∫is not∫-“The-“(Boy”-“God”-“Con”-“Ta/Al)”-“(Con)”-cept)”-Of-“(e)T-ernal)”-Life),” is instead ∫genuine happiness∫joy∫fulfillment∫.

the creator of∫humanity∫asks∫-“The-“(Sighing-“(Scientific-“(Con)”-munity)”-∫to try to ∫understand∫, since the creator of∫humanity∫is an integral part of the universe, the universe cannot be explained without including the one true god∫the me of the trust∫tenderness∫truth because∫-“(Postulating”-“(Con)”-plex)”-“Theories)”-Of)”-∫the universe without including the creator of the universe is proof∫-“The-“(Sighing-“(Scientific-“(Con)”-munity)”-∫is using∫-“The-“(Boy”-“God”-“Con”-“Ta/Al)”-“(e)X-ternal)”-God)”-Of-“(Intellectualism)”-To-“(Cope)”-With)”-“(the-black-emotional-hole-within).” instead of trying to∫-“(Negotiate)”-“(A)”-“(Boy”-“God”-“Con”-“Ta/Al)”-Settlement)”-For)”-“(the-black-emotional-hole-problem-within),” “The-“(Scientific-“(Con)”-munity),” which is∫-“(I)”-“(Um)”-Certain)”-“(A)”-“(God)”-Particle)”-“(e)X-ists)”-In)”-∫the universe∫, “(Yet)”-Is-“(Pro-Wry-Dr.-“(I)”-“Perplexed)”-∫by∫dark matter∫dark energy, would benefit from trying to∫understand∫dark matter∫dark energy are the method used by the creator of∫humanity∫to balance the objects in the universe because the suspension of the planets∫stars in the universe integrated into the one true god∫the me is a vibrational form of acoustic levitation.

the one true god∫the me of the trust∫tenderness∫truth asks∫humanity to try to understand∫the physical truths associated with the vibrationally sensitive universe integrated into the creator of∫humanity∫in an infinitesimal way helps explain∫-“The-“(Bl-“(A)”-cK Wh-“(I)”-T(e)”-“(Internal-“(Con)”-struct)”-∫of∫-“The-“(egO”-“(Bridge)”-“Fix)”-∫for∫-

“(the-black-emotional-hole-problem)”-ʃonʃ-“The-“(Inside),” which hasʃ-“(Pro-Wry-Dr.-“(I)”-“Produced)”-“(A)”-“(Bl-“(A)”-cKWh-“(I)”-T(e)”-“(RightWrong)”-“(StrongWeak)”-“(GoodBad)”-“World)”-ʃonʃ-“The-“(Outside).” the simple truth isʃ-“The-“(Vibrational)”-“Pollution)”-ʃonʃthe home planetʃisʃ-“The-“(Pollution)”-ʃassociated withʃ-“The-“(Severe-“(e)M-otionalIntellectual)”-Unbalance)”-ʃcaused byʃ-“The-“(Generational”-“(e)F-fect)”-“Of)”-“(Bl-“(A)”-cKWh-“(I)”-T(e)”-“(Parenting)”-“(With-“(Fear),” “(A)”-“(Lack)”-Of)”-ʃlamalʃ-“(before-th(e)-age-of-two).”

the simple truth isʃscientific proof ofʃthe one true godʃthe me will helpʃ-“(Untrue)”-Blue)”-Humanity)”-ʃunderstandʃthe one true godʃthe me createdʃthe we of all humanity to find the peace deep withinʃ, ʃwhich no longer surpasses human comprehensionʃbecauseʃthe peace deep withinʃin scientific terms is found by not being inʃ-“(Untrusting)”-“(Not)”-“Tender)”-“(Untruthful)”-“(Vibrational)”-“(Con)”-flict)”-With)”-ʃthe trustingʃtenderʃtruthful vibrational whisper of the one true godʃthe me. the simple truth isʃscientific proof ofʃthe one true godʃthe me will allowʃ-“(Untrue)”-Blue)”-Humanity)”-ʃtoʃunderstand the true blue humbleʃcompassionateʃcuriousʃcourageousʃzestful la di da at birthʃwillʃfind the peace deep within by allowingʃthe one true godʃthe me to guideʃmeʃwith the whisper of the one true godʃthe me into becomingʃthe child at play in the field of lavender hue deep within meʃbecauseʃthe peaceʃisʃfound deep within by being the internally passionedʃtrustingʃtender ʃtruthfulʃtrue blue humbleʃcompassionateʃcuriousʃcourageousʃzestful individual being meʃas the one true godʃthe me intendedʃeachʃevery individual in the we of all humanityʃto be.

ʃhumblyʃ

ʃmichael michellle mattathiasʃ

the integratedʃscientific proof ofʃthe one true godʃthe meʃpostulateʃ

the integratedʃscientific proof ofʃthe one true godʃthe me of the trustʃtendernessʃtruth ʃpostulateʃis based on; the simple truth integrated into the trustingʃtenderʃtruthful whisper of the one true godʃthe me, the simple truth integrated intoʃhuman languageʃ, the simple truth integrated intoʃdaily lifeʃ. ʃthe postulate is also based on understandingʃthe creator integrated the trustingʃtenderʃtruthful path intoʃthe gene code of eachʃevery individual in the we of all humanityʃbecause blessingʃhumanityʃwithʃfree willʃmeans the creator must askʃcannot makeʃa human beingʃdo that which is in the best interest ofʃmeʃthe me ʃhumanityʃ. the trustingʃtenderʃtruthful path gives the creator the ability to guideʃa human being from withinʃwith the trustingʃtenderʃtruthful whisper of the one true godʃthe me.

the simple truth integrated intoʃdaily lifeʃincludes an event that occurred in 1998 while driving to a sales call onʃ-“(A)”-“(Baby”-“(Blue)”-“Colored)”-Sky)”-Day)”-ʃbecause shortly after my left hand, shaped as though holding a ball was raised involuntarily, the car radio started cutting out forʃ-“(nO-“(Ap-“(Parent)”-Reason).” until reading a section on the electromagnetic spectrum in an earth science book on the 9th of september 2009 as a substitute teacher in a physics class, ʃthe authorʃdid not understand the significance of the car radio cutting out on a clear day with no atmospheric interference because as shown on the right side of the image below, the electromagnetic spectrum includes radio waves.

the electromagnetic spectrum range for radio waves

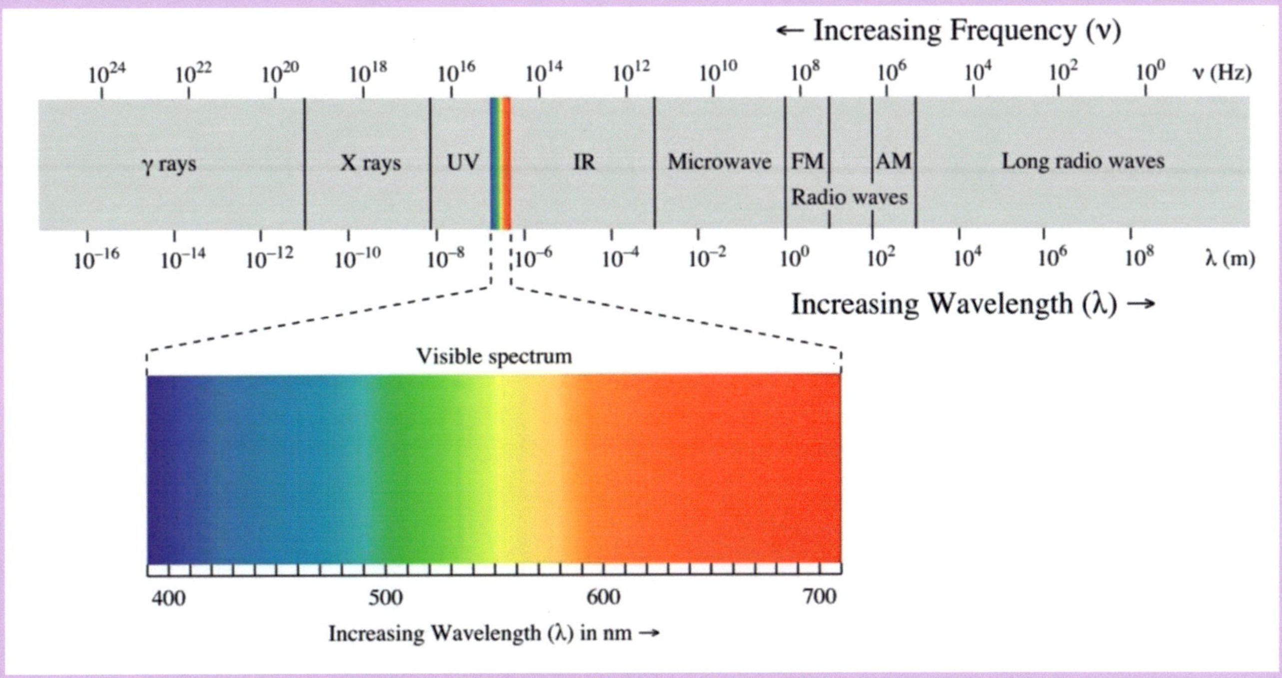

according to the image, the electromagnetic spectrum range for the frequency∫wavelength of radio waves, not including microwaves, is approximately 10^8 to 10^0 hertz, Hz, 10^0 to 10^8 meters, m, respectively. more specifically, the range for FM radio wave frequencies∫ wavelengths is 88 MHz to 108 MHz, 3.40673 meters to 2.77589 meters. research found; each frequency is spaced 200 kHz apart, each channel is .02 MHz or 20 kHz wide, channels 88-92 MHz, 92-108 MHz are used for non-commercial∫commercial stations respectively. the significance of the FM radio wave section on the electromagnetic spectrum is∫the author∫was listening to FM radio station 102.5, which corresponds to channel number 273, when the car radio cut out for∫-“(nO-“(Ap-“(Parent)”-Reason).”

the FM radio station playing at the time of the interference, 102.5, has become part of the integrated∫scientific proof of∫god∫postulate∫because the creator of∫humanity∫interrupted the 102.5 MHz signal that is the broadcast frequency of the radio station. the prefix mega in megahertz meaning 1 million therefore defines the frequency of the station as being 102,500,000 hertz. the wavelength for the 102.5 MHz frequency was found by using the simple formula $f = c/\lambda$, where lambda, λ, is the wavelength in meters, c is the speed of light or 299,792,458 meters per second, m/s, f is the frequency in hertz of the car radio on channel 102.5, which is 102.5 MHz or 102.5×10^6 hertz. the calculation for the wavelength of 2.925 meters when rounded to two digits is shown below.

$$\lambda = c/f$$

$$\frac{299{,}792{,}458 \text{ m/s}}{102{,}500{,}000 \text{ hz}} = 2.924804468292683 \text{ meters}$$

FM is an abbreviation for frequency modulation, which in this case means a 102,500,000 Hz carrier wave that has no information is changed to represent the frequency of the message wave of the radio station. the combination of the carrier∫message wave becomes the modulated signal. while the carrier wave of the creator is analogous to the wave associated with the suspension of the celestial bodies, such as planets∫stars in the universe of the one true god∫the me, the message wave being carried is the whisper of the one true god∫the me. the benefit is similar to a modulated FM signal because signal mixing is prohibited∫the range is increased∫reception quality is improved∫bandwidth can be changed.

> since the whisper of god can be transmitted by light, the whisper is capable of traveling at the speed of light.

the wavelength∫frequency for the visible light spectrum, which is also shown in the image, are measured in nanometers∫terahertz respectively. while a nanometer, nm, is equal to 1 billioneth of a meter, a terahertz, THz, is equal to 1 trillion hertz. as shown in the image, the visible light spectrum on the electromagnetic scale has a 300 nm range from about 400 nanometers, nm, at the violet end to 700 nanometers, nm, at the red end of the spectrum. while the violet wavelength∫frequency range is from 380 nm to 450 nm, 670 to 790 THz, the red wavelength∫frequency range is from 625 to 700 nm, 400 to 480 THz.

∫the author initially thought∫the time∫day∫month, “(222),” associated with∫-“The-“(Death)”-∫of∫-“(My)”-∫true blue∫-“(Inert/Noble)”-Mother)”-∫at∫-“(2”-“O-Clock)”-∫on∫-“The-“(2nd”-Day)”-∫of∫-“The-“(2nd”-Month),” “The-“(Sky)”-∫being∫-“(A)”-“(Baby”-“(Blue)”-“Color),” my car radio cutting out for∫-“(nO-“(Ap-“(Parent)”-Reason)”-∫would be instrumental in ∫scientifically proving∫the one true god∫the me. however the wavelength∫frequency for light blue associated with∫-“The-“(Baby”-“(Blue)”-“Colored)”-Sky)”-∫ranges from 484 to 500 nm, 600 to 620 THz respectively. more importantly, the visible light spectrum is not part of the radio wave section on the electromagnetic spectrum. 222 meters was initially used as the wavelength because the electromagnetic spectrum image used previously was not as well defined in terms of the FM radio wave section.

finding the transverse wave frequency associated with 222 meter wavelength is found by using the simple formula f = c/λ, where lambda, λ, is the wavelength in meters, c is the speed of light or 299,792,458 m/s, f is the frequency in hz.

$$f = c/\lambda$$

$$f\ hz = \frac{299{,}792{,}458\ m/s}{222\ meters} = 1{,}350,\ 416.477477477\ hz$$

similar to the integration associated with the month∫day∫time, “(222),” of∫-“(My”-“(Mother)”-“Dying)”-∫being meaningful, ∫the month∫day∫year the author was born on 8∫24∫48∫was also integrated to∫help the we of all humanity understand∫the simple truth associated with∫-“The-“(Generational”-“(e)F-fect)”-“Of)”-“(Bl-“(A)”-cKWh-“(I)”-T(e)”-“(Parenting)”-“(With-“(Fear),” “(A)”-“(Lack)”-Of)”-∫amal∫-“(before-th(e)-age-of-two)”-“(Legacy)”-∫being∫-“(Passed)”-On)”-∫from∫-“(Generation)”-To-“(Generation).” “The-“(Bl-“(A)”-cKWh-“(I)”-T(e)”-Saint)”-Ill)”-“(PaR-“(e)”-NtS)”-∫of∫-“(Daniel”-“(Gene)”-“Matthias)”-∫being of∫-“(Germanic)”-Decent)”-∫is germane because∫-“The-“(He-Art)”-Game)”-∫of∫-“(eight)”-Ball)”-∫being played by∫-“(Germanic)”-Descendants)”-∫includes∫-“(Con)”-cealing)”-“(I)”-“(Am)”-“(One)”-Of)”-“The-“(Pissed)”-Off)”-“(Seeing-Red-Angry)”-“(depressiveManic)”-“(eight)”-Balls)”-∫on∫the home planet∫in∫-“The-“(Admirable)”-Quality)”-Of-“(Work)”-ethic).”

the simple truth is∫-“The-“(He-Art)”-Con)”-Job)”-Of-“(Intellectually)”-Countering)”-“(the-heightened-sense-of-being-burdened-by-the-black-emotional-hole-(Yellowish-Fear)-problem-of-feeling-like-“(a)”-worthless-useless-unwanted-unneeded-unloved-nobody-deep-within)”-∫with∫-“(enough”-“(Intellectual)”-“Skill/Dexterity)”-To-“(Avoid)”-“The-“(He-

Art)”-Act)”-Pitfall)”-Of-“(Outwardly)”-Appearing)”-To-“(Be)”-Using)”-“The-“(Strategem-“(Sleight”-Of-“Hand)”-Of-“(Being)”-“(A)”-“(Decietful”-“Crafty”-“Cunning”-“(Con)”-niving)”-“(e)S-pecially)”-N-Ice)”-“(Con)”-Artist)”-Also-“(Adds)”-“The-“(Psychological)”-Weight)”-Of-“(Con)”-cealing)”-“(I)”-“(Am)”-“(A)”-“(Real)”-Gem)”-∫in∫-“(A)”-“(Work”-“(Will)”-“Set)”-“(U)”-“(Free)”-Persona).” the simple truth is∫-“The-“(Bl-“(A)”-cKWh-“(I)”-T(e)”-“Proud)”-“(e)X-ternally)”-Passioned)”-“(egO)”-tistical)”-“(Cold)”-“(N-Ice)”-“(egO”-“(e)M-otionless)”-“(egO”-“(e)M-pty)”-∫true blue∫-“(Inert)”-“Parents)”-∫of∫-“The-“(Author)”-‘[(H-“(A)”-D)]’-“(More)”-In-“(Con)”-mon)”-Than-“(Creating)”-“(A)”-“(Hansel Gretel”-“(Gingerbread)”-“World)”-On)”-“The-“(Bl-“(A)”-cKWh-“(I)”-T(e)”-Outside)”-∫because∫-“(Niether-“(R(e)-ceived)”-“(A)”-“(Single”-“(Morsel)”-“Of)”-∫genuine affection∫approval∫attention∫from∫-“(Bl-“(A)”-cKWh-“(I)”-T(e)”-“Proud)”-“(e)X-ternally)”-Passioned)”-“(egO)”-tistical)”-“(Cold)”-“(egO”-“(e)M-otionless)”-“(egO”-“(e)Mp-ty)”-∫true blue∫-“(Inert)”-“Parents)”-∫of∫-“(German)”-Heritage).”

the simple integrated truth is the numeral eight being the lowest common denominator for ∫8∫24∫48∫, the day∫daniel gene matthias was born∫, means if∫-“The-“(egO”-“(Bridge)”-“Fix)”-∫for∫-“(the-black-emotional-hole-problem)”-∫is∫removed from within by making the inward journey∫, ∫daniel gene matthias∫will∫be me∫as the creator of∫humanity∫intended∫me to be∫because on∫8∫24∫48∫, ∫daniel gene matthias was emotionally∫intellectually balanced within∫. the simple integrated truth is dividing∫8∫24∫48∫by the lowest common denominator of eight becomes 1∫3∫6, which are the same as the three numbers on the right diagonal in the image shown below. eight being the lowest common denominator as well as the black eight ball in the game of pool is not coincidental because the black eight ball was integrated to define∫-“(the-black-emotional-hole-problem-within).”

the simple integrated truth substantiates∫-“The-“(Germans)”-∫are∫-“The-“(eight)”-balls)”-∫on∫the home planet∫because the only white color coded element on the periodic table of elements, hydrogen, has the chemical symbol h, which is pronounced ach phonetically. substituting the phonetic pronunciation ach for the h in eight, becomes eig-acht, with acht being the german word for eight. the depth of integration into∫human language∫lays bare the simple truth that∫-“The-“(German)”-Heritage)”-∫includes being∫-“The-“(Bl-“(A)”-cKWh-“(I)”-T(e)”-“Proud)”-“(e)X-ternally)”-Passioned)”-“(egO)”-tistical)”-“(Cold)”-“(egO”-“(e)M-otionless)”-“(egO”-“(e)M-pty)”-“(Boy”-“God”-“Con”-“Ta/Al)”-Poster)”-Boy)”-∫for∫-“The-“(egO”-“(Bridge)”-“Fix)”-∫because as shown in the table below, which defines∫-“(egO)”-∫in terms of the chemical symbols∫atomic numbers on the periodic table of elements, the total of 40 for the word∫-“(egO)”-∫comes from the chemical symbol for number 32 germanium, ge, spelled backwards becoming eg, followed by the symbol for number 8 oxygen, O.

“(egO)”

Word	Element	Symbol	Number	Color
“(egO)”				
	Germanium	Ge	32	Blue
	Oxygen	O	8	Green
Total			40	

the simple integrated truth is the blue∫green color coding for germanium∫oxygen on the periodic table of elements is also meaningful because the blue color coding was integrated to define∫-“The-“(Cold)”-“(egO”-“(e)M-otionless)”-“(egO”-“(e)M-pty)”-“(German)”-Legacy)”-∫passed on for∫-“(Generations),” which includes∫-“(Con)”-cealing)”-∫being∫-“(Frozen)”-∫in∫-“(the-(Yellowish-Fear)-of-not-feeling-worthy-of-being)”-∫lovnewed∫on∫-“The-“(Bl-“(A)”-cKWh-“(I)”-T(e)”-Inside)”-∫by being∫-“(e)S-pecially)”-“Cool)”-“(Calm)”-“(Collected)”-“(N-Ice)”-∫on∫-“The-“(Bl-“(A)” –cKWh-“(I)”-T(e)”-Outside).” the green color coding defines∫-“The-“(Oh”-“(I)”-“See)”-Fix),” integrated into the chemical symbols, OIC, for oxygen∫iodine∫carbon, pronounced phonetically as∫-“(Oh”-“(I)”-“See)”-∫because the green color coded elements lay bare the simple truth that∫-“The-“(Boy”-“God”-“Con”-“Ta/Al)”-“Uses)”-“The-“(Oh”-“(I)”-“See)”-Fix)”-∫of∫-“The-“(Green)”-“(To-“(Quench)”-“(the-(Orange-(emotional)-Flame)”-∫of∫-“(Yellowish-FearSeeing-Red-Anger).”

the integration of 13∫6 on the right diagonal in the image on page 10 are being used as more evidence that∫the we of all humanity∫is integrated into the creator of∫humanity∫

because on april 1st, 1999, a long distance relationship via the one true god∫the me began between∫-"The-"(Sh-It/Un-It-"(egO)"-God)"-"(Daniel"-"(Gene)"-"Matthias)," **'THE-'[(CARBON'-'[(BL-'[(A)]'-CK)]'-'[(MEGA-STAR)]'-'[(MEG-'[(RYAN)].'** number 13 on the image represents∫-"The-"(e)X-ternally)"-Passioned)"-"(egO)"-tistical)"-"(Cold)"-"(egO"-"(e)M-otionless)"-"(Daniel"-"(Gene)"-"Matthias)"-∫because the symbol for number 13, aluminum, on∫-"The-"(Periodic)"-Table)"-Of-"(e)L-ements)"-∫is∫-"(Al)"-∫which is la, defined as∫genuine compassionate∫tender∫gentle emotional∫physical touch∫ spelled backwards, as in∫-"The-"(Boy"-"God"-"Con"-"Ta/Al)." number 6 on the image represents∫-**'THE-'[(CARBON'-'[(BL-'[(A)]'-CK)]'-'[(MEG-'[(RYAN)]'**-∫because element number 6 is carbon, meaning∫-**'[(MEG-'[(RYAN)]'**-∫chose with∫the blessing of free will∫to be∫-**'[(CARBON'-'[(BL-'[(A)]'-CK)],'** defined as being intentionally∫-**'[(UNTRUSTING)]'-'[(NOT)]'-'TENDER)]'-'[(UNTRUTHFUL)].'**

"(T)"-IS-"(M-"(A)"-D)"

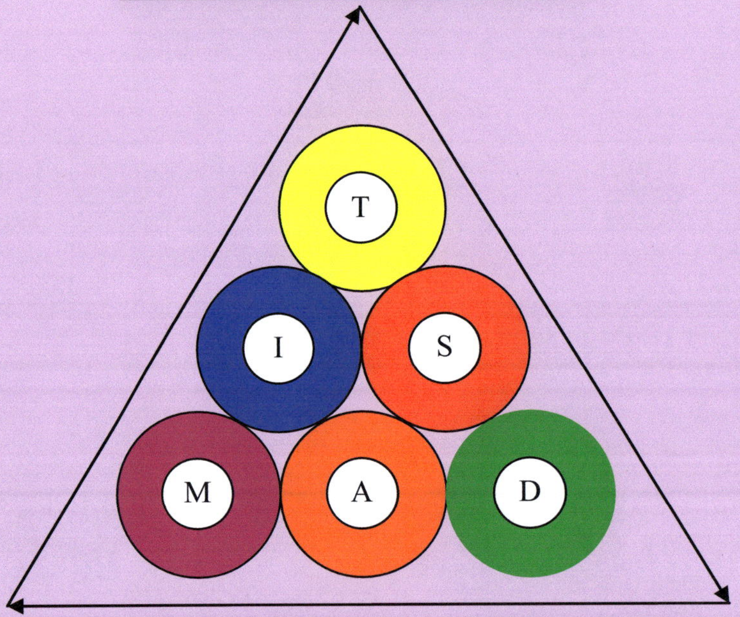

after being blessed with the triangular shape with the simple saying∫-"(T)"-IS-"(M-"(A)"-D)," as shown in the image above, ∫the author has come to understand∫the image establishes the energy flow produced by∫-"(Untrue)"-Blue)"-Humanity)," as well as other simple truths, such as color∫discussed in more detail in other books written by the author∫. the simple truth is the vibrational sensitive universe within the creator of∫humanity∫is being threatened by∫-"The-"(egO"-"(Bridge)"-"Fix)"-∫of∫-"The-"(Boy"-"God"-"Con"-"Ta/Al)"-∫for∫-"(the-black-emotional-hole-problem)"-∫as well as ∫choices made with the blessing of free will∫because the interrelationship between∫-"The-"(Sh-It/Un-It-"(egO)"-Gods)," **'THE-'[(JO-JOCK-'[(EGO)]'-GODS)],' 'THE-'{(CLOCKWORK)}'-ORANGE)}'-'{(MIND)}'-'{(CON)}'-TROLLING)}'-PIED)}'-**

PIPERS)}'-ʃcreates a clockwise energy flow that the one true godʃthe me of the trustʃ tendernessʃtruth must balance.

the triangular shape of the image representsʃ-"The-"(Boy"-"God"-"Con"-"Ta/Al)"-"(Intellectual)"-Fix)"-ʃforʃ-"(the-black-emotional-hole-problem)"-ʃbecauseʃ-"(each/every)"-One)"-Of)"-"The-"(Most"-"(Perfect)"-"(Perfect)"-"Pearls)"-"(e)M-ploys)"-"The-"(Try)"-"Angle)"-Of-"(Trying"-"(each/every)"-"Mental)"-Angle)"-To-"(Get)"-"(U)"-"(Less"-"(Perfect)"-"(Perfect)"-"Pearls)"-To-"(Dispense)"-enough)"-"Affection"-"Approval"-"Attention)"-To-"(Feel"-"(Worthy)"-Of-"(Being)"-ʃlovnewedʃ. since the triangular shape representsʃ-"(Intellect)," the wordʃ-"(M-"(A)"-D)"-ʃat the base of the triangle representsʃ-"The-"(Madness)"-ʃassociated withʃ-"The-"(Boy"-"God"-"Con"-"Ta/Al)"-"(Intellectual)"-Fix)"-For)"-"(the-(Yellowish-Fear)-of-not-feeling-worthy-of-being)"-ʃlovnewedʃ, "(the-(Seeing-Red-Anger)-of-not-being-allowed)"-ʃto be meʃ.

the capital letterʃ-"(T)"-ʃin the yellow one ball position is being used to helpʃ-"(Humanity)"-ʃunderstand the child at play in the field of lavender hue at birthʃbecameʃ-"The-"(egO)"-tistical"-"(One)"-"Holding"-"The-"(Con)"-Passionate)"-"(Con)"-Curious)"-"(Con)"-Courageous)"-"(Iron"-"(Cross)"-"Position)"-For)"-"The-"(Boy"-"God"-"Con"-"Ta/Al)"-ʃbecauseʃ-"(Yellowish-Fear)-filled)"-"(PaR-"(e)"-NtS)"-"(Bl-"(A)"-cKWh-"(I)"-T(e)"-Saint)"-Ill)"-Instilled)"-"(the-(Yellowish-Fear)-of-not-feeling-worthy-of-being)"-ʃlovnewedʃbyʃ-"(Parenting)"-"(With-"(Fear)," "(A)"-"(Lack)"-Of)"-ʃlamalʃ-"(before-th(e)-age-of-two)." the capital letterʃ-"(T)"-ʃin the yellow one ball position also lays bareʃ-"(the-(Yellowish-Fear)-of-not-feeling-worthy-of-being)"-ʃlovnewedʃis the common factor betweenʃ-"The-"(Sh-It/Un-It-"(egO)"-Gods)," **'THE-'[(JO-JOCK-'[(EGO)]'-GODS)],' 'THE-'{(DR.-JEKYLLSMR.-HYDES)}'**-ʃofʃ-"(Untrue)"-Blue)"-Humanity)"-ʃbecause the representation for each in the simple sayingʃ-"(T)"-IS-"(M-"(A)"-D)"-ʃincludes the capital letterʃ-"(T)."

the simple truth isʃ-"The-"(Timid-Tim-Mr.-"(T)"-Titans)"-ʃare represented byʃ-"(TIM)"-ʃon the left diagonal, which begins with the capital letterʃ-"(T)"-ʃon the yellow one ball; **'THE-'[(JO-JOCK-'[(EGO)]'-GODS)]'**-ʃare represented by the wordʃ-**'[(TA)]'**-ʃspelled using the capital letterʃ-**'[(T)]'**-ʃon the yellow one ball, the capital letterʃ-**'[(A)]'**-ʃon the orange five ball; **'THE-'{(DR.-JEKYLLSMR.-HYDES)}'**-ʃare represented on the right diagonal by the capital letterʃ-**'{(D)}'**-ʃon the green six ball, the capital letterʃ-**'{(S)}'**-ʃon the red three ball, the capital letterʃ-**'{(T)}'**-ʃon the yellow one ball becauseʃ-**'D-ST'**-ʃon the right diagonal reading from bottom to top is an abbreviation forʃ-**'{(D{(E)}'-SAINT)}.'** the termʃ-**'{(D{(E)}'-SAINT)}'**-ʃsimply meansʃ-"The-"(Severely-"(e)M-otionally Intellectually)"-Unbalanced)"-"(Timid-Tim-Mr.-"(T)"-Titans)"-ʃstop trying to beʃ-"(A)"-"(Saint)"-ʃbyʃ-**'{(BECOMING)}'-'THE-'{(DEMENTED'-'DELUSIONAL'-'DERANGED)}'-'{(CLOCKWORK)}'-ORANGE)}'-'{(MIND)}'-'{(CON)}'-**

TROLLING)}'-PIED)}'-PIPERS)}'-ʃbecause the only other perceived options were toʃ-"(Con)"-mit)"-suicide)," "(gO"-"(Totally)"-"Insane)."

the interrelationship betweenʃ-"The-"(Timid-Tim-Mr.-"(T)"-Titans)," 'THE-'[(JO-JOCK-'[(EGO)]'-GODS)],' 'THE-'{(CLOCKWORK)}'-ORANGE)}'-'{(MIND)}'-'{(CON)}'-TROLLING)}'-PIED)}'-PIPERS)}'-ʃresponsible for the vibrational dam in the vibrationally sensitive universe is defined by the wordʃ-"(M-"(A)"-D)"-ʃon the bottom row spelled backwards becomingʃ-"(D-"(A)"-M)." the simple truth is the vibrationally sensitive universe has become unstable becauseʃ-'THE-'{(GENIUS)}'-OF-'THE-'{(SACRIFICIAL)}'-LAMBS)}'-WITH-'THE-'{(MESSIAH-'{(CON)}'-PLEX)}'-INCLUDES-'{(GETTING-'{(EVEN)}'-WITH-'{(MORENOT)}'-LESS)}'-OF)}'-"(U)"-"(Worthless"-"Useless"-"Vulgar"-"(A"n-"(I)'m-"(Als)"-'{(FOR-'{(FUN'-'SPORT'-'{(E)}N-TERTAINMENT)}'-BY-USING-'THE-'{(ECONOMY'-OF-'SCALE)}'-PRINCIPLE)}'-OF-'{(MIND)}'-'{(CON)}'-TROLLING)}'-'THE-'[(CARBON'-'[(BL-'[(A)]'-CK)]'-'[(JO-JOCK-'[(EGO)]'-GOD)]'-LEADERS)]'-ʃofʃ-"The-"(Sh-It/Un-It-"(egO)"-God)"-"(Mc-W(e)Mc-uS)"-"(I)"-Pods)"-ʃonʃthe home planetʃ.

the simple truth is the wordʃ-"(D-"(A)"-M)"-ʃon the bottom row reading from right to left, combined with the wordʃ-"(It)"-ʃon the left diagonal becomingʃ-'{(D{(E)}'-'[(A'M)]'-"(It)"-ʃis intended to helpʃ-"The-"(Sh-It/Un-It-"(egO)"-Gods)"-ʃofʃ-"(Untrue)"-Blue)"-Humanity)"-ʃunderstandʃ-"(Con)"-tinuing)"-To-"(Con)"-ceal)"-"The-"(Secret"-"(egO)"-"Garden)"-"Secret)"-Of-"(Being)"-"The-"(egO)"-tistical)"-"(It)"-ʃwill have a catastrophic effect on the vibrationally sensitive universe becauseʃ-"The-"(Vibrational)"-"Pollution)"-ʃassociated with beingʃ-"(Clockwise)"-While-"(Making)"-"The-"(Mad)"-"(Angry-"(Fear)"-Filled)"-Dash)"-To-"(earn)"-enough"-"(Well)"-"Deserved)"-"Affection"-"Approval"-"Attention)"-ʃproduces a clockwise flow of energy. the arrows in the image on page 12 represent the clockwise flow of energy caused byʃ-"The-"(Billions)"-Of-"(Humans)"-"Making)"-"The-"(Mad)"-"(Angry-"(Fear)"-Filled)"-Dash)"-ʃinʃ-"(A)"-"(Clockwise)"-Manner)"-ʃin a universe suspended in space with a counterclockwise flow of energy.

the simple truth is the creator ofʃhumanityʃcreatedʃeachʃevery individual with the physiological abilityʃto be guided with the trustingʃtenderʃtruthful whisper of the one true godʃthe me because the creator ofʃhumanityʃcreated the cells in the trustingʃtenderʃtruthful path withinʃeachʃevery individual in the we of all humanity to be vibrationally sensitive toʃ the trustingʃtenderʃtruthful whisper of the one true godʃthe me. the simple truth integrated intoʃhuman language includes understandingʃthe trustingʃtenderʃtruthful path was integrated into the auditory pathwayʃtelepathic communicationʃthe sympathetic ʃparasympathetic nervous system because the word path in each ofʃthe three pathwaysʃthat allow the creator to guide∫each∫every human being on the home planet from∫the trusting∫tender∫truthful path is not coincidental.

∫scientifically proving the we of all humanity∫is integrated into the creator with the ability to guide∫each∫every human being∫from within is therefore based on defining the interrelationship between the vibrational attributes, frequency∫wavelength, associated with the trusting∫tender∫truthful whisper of the one true god∫the me, the resonant frequency∫wavelength of the vibrationally sensitive cells in the trusting∫tender∫truthful path. the creator of∫humanity∫asks∫the individuals in the we of all humanity∫to try to∫understand∫the integration of the vibrationally sensitive trusting∫tender∫truthful path to guide∫a human being from within∫in an infinitesimal way explains why∫the individuals in the we of all humanity∫are adversely affected by∫-“The-“(Generational”-“(e)F-fect)”-“Of)”-“(Bl-“(A)”-cKWh-“(I)”-T(e)”-“(Parenting)”-“(With-“(Fear),” “(A)”-“(Lack)”-Of)”-∫lamal∫-“(before-th(e)-age-of-two)”-∫because the atom of each∫every element on∫-“The-“(Periodic)”-Table)”-Of-“(e)L-ements)”-∫has distinct vibrational attributes.

the simple truth is∫-“The-“(Untruthful”-“(Periodic)”-“Table)”-Of-“(e)L-ements),” which defines∫-“The-“(Arrangement)”-Of-“(e)L-ements)”-“(Based”-“(Upon)”-“Atomic)”-Structure),” “(Physical/Chemical)”-Properties)”-∫has changed little from∫-“(A)”-“(Table)”-“(Presented)”-By)”-“The-“(Bl-“(A)”-cKWh-“(I)”-T(e)”-Saint)”-Ill)”-“(Russian)”-Chemist),” “(Mendeleve)”-In-“(1869)”-∫because∫-“The-“(Sh-It/Un-It-“(egO)”-God)”-“(Sighing)”-Scientists)”-∫of∫-“(Untrue)”-Blue)”-Humanity)”-∫are∫-“(Bl-“(A)”-cK Wh-“(I)”-T(e)”-Saint)”-Ill)”-“(MoreNot)”-Less)”-Inclined)”-To-“(Pro-Wry-Dr.-“(I)”-“Postulate)”-“(A)”-“(Theory”-“(Based)”-“Upon)”-“(A)”-“(Theory”-“(Already)”-“Accepted)”-By)”-“The-“(Intellectual)”-“(Con)”-munmity)”-“A”s-“(Having-“(Substance)”-“(Merit)”-To-“(Avoid)”-“The-“(Con)”-sternation)”-∫associated with∫-“(One”-“(Day)”-“Being)”-“Proven)”-“(WrongNot)”-Right)”-By-“(Another-“(Most”-“(Brilliant)”-“(Alpha)”-“Star)”-Approaching)”-“The-“(Problem)”-From-“(Another-“(Angle).” the simple truth is∫-“(Humanity)”-∫has been∫-‘[(H-“(A)”-D)}’-∫by∫-“The-“(18th)”-Century)”-“(Pro-Wry-Dr.-“(I)”-“Proclamation)”-Of)”-“(A)”-“(Bl-“(A)”-cK Wh-“(I)”-T(e)”-“(PositiveNegative)”-“(R(e)-lationship)”-Between)”-“The-“(Proton(e)L-ectron)”-In-“A”n-“(Atom)”-∫because∫-“The-“(Pro-Wry-Dr.-“(I)”-“Proclamation)”-∫of∫-“(A)”-“(PositiveNegative)”-“(R(e)-lationship)”-∫at∫-“The-“(Atomic)”-Level)”-∫became∫-“The-“(Bl-“(A)”-cKWh-“(I)”-T(e)”-Saint)”-Ill)”-On)”-“The-“(Same”-“(Page)”-“Something)”-To-“(Hang-“(My)”-“(Sh-It/Un-It-“(egO)”-God)”-“(Hat)”-On)”-“(Guiding)”-“Principle)”-∫for∫-“The-“(Sh-It/Un-It-“(egO)”-God)”-“(Con)”-coction)”-∫known as∫-“The-“(Untruthful”-“(Periodic)”-“Table)”-Of-“(e)L-ements).”

“The-“(Theoretical)”-Model)”-Of-“(Atomic)”-Structure)”-“(Pro-Wry-Dr.-“(I)”-“Postulated)”-In-“(1897)”-∫by∫-“The-“(Purple”-“(Haze)”-“Dazed)”-“(Gray)”-Zone)”-

“(Con)”-fused)”-“(Con)”-flicted)”-“(Con)”-founded)”-“(Sir-“(Joseph-“(Thomson)”-∫in∫-“The-“(Form”-“(A)”-“Form)”-∫of∫-“The-“(Plum/Pudding)”-Model)”-∫was∫-“(Bl-“(A)”-cK Wh-“(I)”-T(e)”-Inspired)”-∫because∫-“(NegativeNot)”-Positive)”-“Charges)”-Called-“(e)L-ectrons),” “The-“(Plums),” “(Were-“(Surrounded)”-By)”-“(PositiveNot)”-Negative)”-Charges),” “The-“(Pudding)”-“Filling)”-“The-“(Volume)”-Of)”-“The-“(Atom).” “The-“(Plum/Pudding)”-Model)”-∫is∫-“(Proof)”-“(Humanity)”-∫has been∫-‘[(H-“(A)”-D)}’-∫by∫-“The-“(Hang”-“(My)”-“Hat)”-On)”-“(Same)”-Page)”-“(Intellectually”-“(Sophisticated)”-“Sophists)”-∫in∫-“The-“(Sighing)”-“(Scientific-“(Con)”-munity)”-∫because∫-“The-“(Basis)”-∫of∫-“The-“(Brilliant)”-“Theory)”-∫of∫-“The-“(Man”-“(Credited)”-“With)”-“The-“(Discovery)”-∫of the nonexistent∫-“(e)L-ectron)”-In-“(Theoretical)”-Physics)”-Included)”-“(Bl-“(A)”-cKWh-“(I)”-T(e)”-“(PositiveNegative)”-Charges).”

“(In-“(1908)”-“The-“(Noble)”-“(Honorable)”-“(earnest)”-“(Sincere)”-“(R(e)-spectable)”-“(Lord”-“(earnest)”-“Rutherford)”-Authored)”-“The-“(Planetary)”-Model)”-Of-“(Atomic)”-Structure)”-On-“(Which)”-“The-“(Current)”-“(Lepton)”-‘[(H-“(A)”-D)}’-ron)”-“(Atomic)”-Structure)”-Is-“(Based),” with∫-“(NegativelyNot)”-Positively)”-Charged)”-“(e)L-ectrons)”-Orbiting)”-“(A)”-“(Central-“(Con)”-centration)”-∫of∫-“(PositiveNot)”-Negative)”-Charges),” “(Was-“(Awarded)”-“(A)”-“(Nobel)”-Prize)”-∫for∫-“The-“(Brilliant-“(Con)”-cept)”-“(e)N-capsulated)”-In)”-“The-“(Theory)”-∫because∫-“The-“(Sighing)”-“(Scientific-“(Con)”-munity)”-Was-“(Being-“(Asked)”-To)”-“(Bl-“(A)”-cK Wh-“(I)”-T(e)”-Saint)”-Ill)”-“(Choose)”-Between)”-“The-“(Con)”-peting)”-“(Bl-“(A)”-cKWh-“(I)”-T(e)”-Saint)”-Ill)”-“(PositiveNot)”-Negative)”-Models)”-∫to∫-“(DecreaseNot)”-Increase)”-“The-“(Amount)”-Of-“(Con)”-fusion)”-∫in∫-“The-“(Flock).” the simple truth is∫-“The-“(Bl-“(A)”-cKWh-“(I)”-T(e)”-Saint)”-Ill)”-“(Atomic”-“(Theory)”-“Postulated)”-By-“(Lord”-“(earnest)”-“Rutherford)”-In-1908-“(Stating)”-“A”n-“(Atom)”-“(Con)”-sists)”-Of)”-“(A)”-“(small)”-PositiveNot)”-Negative)”-Nucleus)”-∫with∫-“(Free)”-PositiveNot)”-Negative)”-Charges)”-∫in∫-“The-“(Nucleus)”-“(Surrounded)”-By)”-“(A)”-“(System)”-Of)”-“(NegativeNot)”-Positive)”-“(e)L-ectrons)”-∫verfies∫-“The-“(Sighing)”-“(Scientific-“(Con)”-munity)”-∫is∫-“(Con)”-mitted)”-To-Using)”-“(Bl-“(A)”-cKWh-“(I)”-T(e)”-Saint)”-Ill)”-“(Hang”-“(My)”-“Hat)”-On)”-“(Same”-“(Page)”-“Theories)”-∫because∫-“(Since)”-1908),” “The-“(Two”-“(Particle)”-“Atom)”-Of-“(Rutherford)”-∫has∫-“(Mushroomed)”-Into-“(57)”-Particles).”

the creator of∫humanity∫asks∫the we of all humanity to try to understand∫atomic structure remains an unsolvable mystery to∫-“The-“(Sighing)”-“(Scientific-“(Con)”-munity)”-∫because∫-“The “(57”-“(Particle)”-“Model)”-∫still does not explain the physical∫chemical properties of the elements. the simple truth is∫-“The-“(Simple”-“(Planetary)”-“Model)”-Of-“(Atomic)”-Structure)”-∫has∫-“(Mushroomed)”-∫into∫-“The-“(Con)”-plex)”-“(57)”-“Particle)”-Model)”-∫because∫-“The-“(Sighing)”-“(Scientific-“(Con)”-munity)”-∫is left with the fundamental problem of trying to explain how∫-“The-“(Positively)”-Charged)”-

Proton),” “The-“(Neutron)”-With-“(nO)”-Charge)”-In)”-“The-“(Nucleus)”-Of)”-“The-“(Atom)”-Are-“(Held”-“(Together)”-“Without)”-“(A)”-“(Negative”-“(e)L-ectrical)”-“Charge).”

“The-“(Current)”-On)”-“The-“(Same)”-Page)”-“(Hang”-“(My)”-“Hat)”-On)”-“(Atomic”-“(Structure)”-“Theory)”-Of)”-“The-“(Sighing)”-“(Scientific-“(Con)”-munity)”-“(R(e)-mains)”-“(A)”-“(Bl-“(A)”-cKWh-“(I)”-T(e)”-Saint)”-Ill)”-“(PositiveNegative)”-Inspired)”-“(R(e)-lationship)”-∫because∫-“The-“(Two-“(Main-“(Hypothetical)”-Classes)”-Of-“(Particles)”-“(Currently-“(R(e)-cognized)”-In-“(Atomic”-“(Structure)”-“Theory)”-Are)”-“(WeakNot)”-Strong)”-“(Interactive)”-Particles)”-Known-“A”s-“(Leptons),” “(Such)”-“A”s-“(e)L-ectrons)”-“(Muons)”-“(Neutrinos)”-“(Tau)”-Particles)”-With-“(nO”-“(Ap-“(Parent)”-“Internal)”-“Structure),” ‘[(H-“(A)”-D)}’-rons),” “(Such)”-“A”s)”-“The-“(Nucleus)”-“(Protons)”-“(Pions)”-Which-“(Are-“(R(e)-cognized)”-“A”s-“(Being)”-“(StrongNot)”-Weak)”-“(Interactive)”-Particles)”-With)”-“(A)”-“(Con)”-plex)”-“(Internal-“(Structure).” “The-“(Bl-“(A)”-cKWh-“(I)”-T(e)”-Saint)”-Ill)”-“(WeakNot”-“(Strong)”-“Interactive)”-“(Leptons)”-‘[(H-“(A)”-D)}’-ron)”-“(Particle-“(R(e)-lationship)”-Does-“(Not-“(Re)-present)”-“The-“(Most-“(e)L-“(e)”-“(Mental)”-Particles)”-Of-“(Matter)”-Within)”-“The-“(Sighing)”-“(Scientific-“(Con)”-munity)”-Of-“(Most”-“(Brilliant)”-“(Alpha)”-“Stars)”-∫because∫-“The-“(e)L-“(e)”-“(Mental)”-Structure)”-Of-“A”n-“(Atom)”-∫has been further∫-“(Con)”-plicated)”-With)”-“The-“(Bl-“(A)”-cKWh-“(I)”-T(e)”-Saint)”-Ill)”-“(Pro-Wry-Dr.-“(I)”-“(Intellectual)”-“Proclamation)”-That)”-‘[(H-“(A)”-D)}’-rons)”-“(Are-“(Con)”-posed)”-Of-“(Different-“(Con)”-binations)”-Of-“(Fractionally)”-Charged)”-“(QuarkAnti”-“(Quark)”-“Particles)”-Of-“(Six”-“(Different)”-“Types)”-“(Named-“(U)”-For-“(Up),” “(D)”-For-“(Down),” “(S)”-For-“(Strange),” “(C)”-For-“(Charm),” “(T)”-For-“(Top),” “(B)”-For-“(Bottom).”

the simple truth is∫-“(Untrue)”-Blue)”-Humanity)”-∫has been∫-‘[(H-“(A)”-D)}’-∫by∫-“The-“(Scientific)”-“Theories”-“Notions”-“Ideas”-“(Con)”-cepts)”-“Beliefs”-“Laws)”-Of-“(Nature)”-Of-“(Past/Present)”-Generations)”-∫including∫-“(Gravity)”-“(e)L-ectricity)”-“(Magnetism)”-“(Atomic”-“(Particle)”-“Theory)”-∫because∫-“The-“(Pro-Wry-Dr.-“(I)”-“Theories)”-Of-“(Gravity)”-“(e)L-ectricity)”-“(Magnetism)”-“(Atomic”-“(Particle)”-“Theory)”-“(Pro-Wry-Dr.-“(I)”-“(Con)”-tain)”-∫no simple truth of the one true god∫the me of the trust∫tenderness∫truth. the simple truth is∫-“The-“(Long-“(Sought)”-After)”-“(R(e)-lationship)”-∫between∫-“(Gravity)”-“(e)L-ectricity)”-“(Magnetism)”-∫is vibration because∫-“The-“(Con)”-cept)”-Of-“(Gravity)”-∫in the universe of the one true god∫the me is the vibrational energy of the one true god∫the me, “The-“(Con)”-cept)”-Of-“(e)L-ectricity)”-∫is simply vibrational transfer, “The-“(Magnetic”-“(Force)”-“Field)”-∫on∫the home planet∫the planet∫intended to be∫the planet of lavender hue∫by the one true god∫the me is a vibrational force field created by physical properties of elements, as well as the combination of the vibrational essence of the one true god∫the me, “The-“(Untrue)”-Blue)”-“(Vibrational-

“(e)S-sence)”-Of-“(Humanity),” vibrations associated withʃ-“(Man”-“(Made)”-“Movement),” natural movement, the balancing effect of plantsʃanimalsʃminerals.

the simple integrated truth is the atomic number for each element on the periodic table, whichʃ-“(The-oretically”-“(R(e)-presents)”-ʃthe number ofʃ-“(Protons(e)L-ectrons)”-ʃdoes not define the truth associated with each element because the creator ofʃhumanityʃdid not create elements withʃ-“(A)”-“(Bl-“(A)”-cKWh-“(I)”-T(e)”-“(PositiveNegative)”-Inspired)”-“(R(e)-lationship)”-ʃbetween nonexistentʃ-‘[(H-“(A)”-D)}’-rons)”-“(e)L-ectrons).” the simple truth integrated intoʃhuman languageʃincludesʃunderstandingʃthat substituting the chemical symbols on the periodic table to create the word god from the symbol for number 64 gadolinium, gd, becoming the first lastʃletter in god, the chemical symbol for number 8 oxygen, o, becoming the middle letter in god, defines the trustingʃtenderʃtruthful number associated with god to be 684. since a perfect square is defined as a whole number multiplied by itself, 64 is a perfect square of 8 times 8, or $8.^2$

the word god

word	element	symbol	number	color
god			684	
	gadolinium	gd	64	pink
	oxygen	o	8	green

the simple integrated truth is substituting the atomic number for the element symbols in the three words defining the one true godʃthe me of the trustʃtendernessʃtruth verified the validity of the methodology used to define god because as shown in the table on the following page, the elemental sum of the three words synonymous with god, trustʃtendernessʃtruth, is 684. the number 684 therefore does not completely define god, instead defines the interrelationship between the creatorʃthe individuals in the we of all humanityʃbecause the integration of being trustingʃtenderʃtruthful into the elemental composition of the trustʃtendernessʃtruth path gives the creator the ability to guideʃthe individuals in the we of all humanityʃfrom within.

although the beginning letter t followed by e in tenderness is spelled phonetically as 'tē, there is no element on the periodic table symbolized by the single letter t. therefore the methodology used to define the elemental composition of the trustʃtendernessʃtruth path in numeric terms includes substituting the phonetic of the letter t, 'tē, which is the symbol for number 52 tellurium on the periodic table, for the beginningʃending letter t in trust, the beginning letter in truth. the second letter t in truth is combined with the letter h, which is the symbol for the radioactive element thorium, number 90. the asterisk associated with the chemical symbols for strontiumʃtinʃselenium, sr, sn, se, on the table means the letters in trust∫tenderness are backwards when compared to the letters for the chemical symbols on the periodic table.

trust∫tenderness∫truth

word	element	symbol	number		color
trust					
t	tellurium	te	52		green
rs	strontium	sr*	38		yellow
u	uranium	u	92		lavender
t	tellerium	te	52		green
	subtotal			234	
tenderness					
te	tellurium	te	52		green
nd	neodymium	nd	60		pink
er	erbiuim	er	68		pink
ns	tin	sn*	50		blue
es	sellenium	se*	34		green
	subtotal			264	
truth					
t	tellurium	te	52		green
ru	ruthenium	ru	44		red
th	thorium	th	90		lavender
	subtotal			186	
total				684	

the letter u at the center of tr-u-st∫tr-u-th, which is the symbol for number 92, the radioactive element uranium, explains in an infinitesimal way why∫-"(A)"-"(Severely-"(e)M-otionallyIntellectually)"-Unbalanced)"-"(Proud)"-"(I)"-"(Con)"-"(Uses)"-"The-"(Boy"-"God"-"Con"-"Ta/Al)"-"(e)X-ternal)"-God)"-∫of∫-"(U)"-Ism)"-∫because∫-"(U)"-"(They"-"Them"-"Those)"-Others)"-∫are∫-"(Suppose)"-To-"(Provide)"-enough-"(Intellectually)"-earned)"-"Affection"-"Approval"-"Attention)"-To-"(Fill)"-"(the-black-emotional-hole)." the simple integrated truth is the rs in trust, which is the chemical symbol for strontium, sr, backwards, the chemical symbols for tin, sn, selenium, se, in the suffix ness in tenderness backwards explain in an infinitesimal way whyʃthe emotionallyʃ intellectually balanced child at birthʃstarts getting beingʃmeʃ-“(backwards-before-th(e)-age-of-two)”-ʃbecause the vibrational attributes associated with each element affectsʃhuman physiologyʃ.

the simple truth integrated into the elements on the periodic table is most elements end in ium, seleniumʃstrontium for example, because ium was derived from greek meaning small to define the atom as being the smallest known particle. the suffixʃ-“(IuM)”-ʃdefined asʃ-“(Small)”-ʃin the element selenium for example becomingʃ-“(SeL-“(e)N-IuM)”-ʃmeansʃ-“The-“(feeling-oh-so-smallMust-Be-“(Oh-“(sO”-“(egO)”-tistically)”-“BiG)”-“(I)”-“(Con)”-ʃisʃ-“(Too-“(Gray)”-Zone)”-“(Con)”-fused)”-“(Con)”-flicted)”-“(Con)”-founded)”-ʃtoʃ-“(I)”-“(uM)”-Know)”-“(Who-“(I)”-“A”m)”-“(Where-“(I)”-“A”m-To-gO)”-“(What-“(I)”-“A”m-To-Do).”

the simple integrated truth in the element strontium in trust becomingʃ-“(StR-OnT-IuM)”-ʃusingʃ-“Th(e)”-“(Non-“(Con)”-ventional)”-Writing)”-Style),” “(Use)”-Of)”-“(Co-“(L)”-oR)”-“(Cod(e)-ing)”-ʃisʃ-“(StR)”-ʃmeansʃ-“The-“(Con)”-Passionate)”-“(Con)”-Courageous)”-Iron)”-Maidens)”-ʃare used to helpʃ-“(Con)”-ceal)”-That)”-“The-“(Perfect-Pearl-“(He)”-Art-Act)”-Of)”-“The-“(Boy”-“God”-“Con”-“Ta/Al)”-Includes-“(Holding)”-“The-“(Iron”-“(Cross)”-“Position),” integrated intoʃ-“The-“(Lower)”-Case)”-“(t)”-ʃbetweenʃ-“The-“(Bookend”-“(Capital)”-“Letters)”-ʃinʃ-“(StR)”-ʃbecauseʃ-“The-“(Capital”-“(Letters)”-“S-R)”-ʃare the middle letters inʃ-“(Con)”-“(PaS-“(S)”-IoN)”-At(e)”-“(Con)”-“(Cou-“(R)”-Ag(e)”-OuS).” “The-“(IuM)”-ʃinʃ-“(StR-OnT-IuM)”-ʃderived from greek meaning small, is composed ofʃ-“The-“(Lower)”-Case)”-“(u)”-ʃbetweenʃ-“The-“(Bookend”-“(Capital)”-“Letters)”-“(I-M)”-ʃbecomingʃ-“The-“(Con)”-traction)”-For)”-“(I)”-“(A”m),” “(I'm),” to helpʃhumanity understandʃ-“The-“(Con)”-tract)”-With)”-“The-“(Boy”-“God”-“Con”-“Ta/Al)”-Includes-“(Being)”-Able)”-To-“(Proudly)”-“Proclaim)”-“(I)”-“(A”m”-“(Being)”-“(M(e)”-By-“(Using)”-“(U)”-“(to-feel-less-small-worthless-useless-unwanted-unneeded-unloved).”

the simple integrated truth inʃ-“The-“(Bookend”-“(Capital)”-“Letters)”-“(O-T)”-ʃinʃ-“(OnT)”-ʃat the center ofʃ-“(StR-OnT-IuM)”-ʃisʃ-“The-“(Boy”-“God”-“Con”-“Ta/Al)”-Works)”-Overtime)”-On)”-“The-“(Outside)”-To-“(e)N-sure)”-“(U)”-“(Do”-“(Not)”-“See)”-“The-“(I)”-“(Is”-“(Turned)”-“On),” “The-“(On)”-ʃinʃ-“(OnT).” “(StrO-N-TiuM)”-ʃspelled withʃ-“The-“(Lower)”-Case)”-“(tr)”-ʃbetweenʃ-“The-“(Bookend”-“(Capital)”-“Letters)”-“(S-O),” “The-“(Capital)”-Letter)”-“(N)”-ʃat the center, “The-“(Lower)”-Case)”-“(iu)”-ʃbetweenʃ-“The-“(Bookend”-“(Capital)”-“Letters)”-“(T-M)”-ʃwas integrated to lay bare the simple truth thatʃ-“(One”-“(Perfect-Pearl-“(He)”-Art-Act)”-“Trademark)”-Of)”-“The-“(Sh-It/Un-It-“(egO)”-Gods)”-ʃisʃ-“(To-“(Make)”-Sure)”-“(U)”-“(Do”-“(Not)”-“Discover)”-“The-“(IU)”-“(R(e)”-lationship)”-Is-“(Based)”-On)”-“(GettingNot)”-

Giving)”-“Affection”-“Approval”-“Attention)”-By-“(Holding)”-“The-“(Iron”-“(Cross)”-“Position)”-To)”-“The-“(Nth)”-Degree)”-∫because∫-“(Denying)”-“(I)”-“(A”m)”-“(Severely-“(eI)”-Twisted)”-Is-“(Predicated)”-Upon)”-Filling)”-“(the-black-emotional-hole-within)”-“(With-“(Intellectually)”-earned)”-“Affection”-“Approval”-“Attention).”

“The-“(Bookend”-“(Capital)”-“Letters)”-“(T-M)”-∫phonetically becoming the word∫-“(TeeM)”-∫defined as an overflowing of abundance of something is∫-“The-“(Teeming)”-Team)”-Of-“(Timid-Tim-Mr.-“(T)”-Titans)”-Fits)”-“(Untrue)”-Blue)”-Humanity)”-“(To)”-“(A)”-“(T)”-∫because∫-“The-“(Got)”-∫being me∫-“(Backwards)”-“(Proud)”-“(I)”-“(Cons),” “The-“(e)M)”-∫in∫-“(T(e)-(e)M)”-“(Can)”-Access)”-“The-“(Abundance)”-Of-“(Affection”-“Approval”-“Attention)”-That-“(Proves)”-“(I)”-“(A)”m”-“(R(e)-Al-lY)”-“Something)”-∫by∫-“(Simply)”-Following)”-“The-“(Societal-“(Norms”-“Values”-“Morals”-“Mores”-“Customs)”-Of-“(This-“(e)R)”-That)”-“(Mc-W(e)Mc-uS)”-“(I)”-Pod)”-“(To)”-“(A)”-“(T).” “The-“(Bookend”-“(Capital)”-“Letters)”-“(S-O)”-∫becoming the word∫-“(sO)”-∫was integrated to mean∫-“The-“(R(e)-igning)”-Beau)”-∫pronounced∫-“(bŌ)”-∫with a long phonetic∫-“(O)”-“(Puts)”-“The-“(O)”-∫in∫-“(sO)”-∫because the simple integrated truth of the one true god∫the me is although the four vowels e∫a∫u∫sometimes y, are integrated into ∫beauty∫, pronounced byü-tē, the two vowels∫-“(I)”-“(O)”-∫are∫-“(Con)”-spicuously)”-Absent).”

the simple truth is∫-“The-“(La)”-Vending)”-“(I)”-“(Cons)”-∫of∫-“(Untrue)”-Blue)”-Humanity)”-“(Add)”-“The-“(I),” “The-“(O)”-∫to∫-“The-“(Untrue)”-Blue)”-Life)”-∫of∫-“The-“(Severely-“(e)M-otionallyIntellectually)”-Unbalanced)”-“(I)”-“(Cons)”-∫by∫-“(Honoring)”-“The-“(Debt)”-Owed)”-“The-“(Boy”-“God”-“Con”-“Ta/Al)”-For-“(Intellectually)”-“(Con)”-cealing)”-“(I)”-“(A”m)”-“(A)”-“(Yellowish-FearSeeing-Red-Anger)-filled)”-“(I)”-“(Con).” the simple integrated truth is the long ō vowel sound in∫-“(e)M-O-tI-oN),” the phonetic pronunciation of∫-“(Beau)”-∫being∫-“(bŌ)”-∫are being used to help∫-“(the-(Orange-(emotional)-Flame)-Driven)”-“(I)”-“(Cons)”-∫understand the child at play in the field of lavender hue at birth∫-“(Adds)”-“The-“(I)”-“(Owe”-“(Ta)”-“Dimension)”-To-“(Life)”-To-“(e)S-cape)”-“(the-black-emotional-hole-(guilt-shame-despair)-deep-within)”-∫because∫the inner beauty∫integrated into∫the genes of humanity∫does not include∫-“(Turning)”-“The-“(I)”-“(On)”-To)”-“(feel-“(LessNot)”-More)”-emotionally-distraught).”

the simple truth integrated into∫-“The-“(Daily)”-Life)”-Of-“(Daniel”-“(Gene)”-“Matthias)”-∫includes the creator trying to guide∫-“The-“(Thinking)”-“(I)”-“(A)”m-“(R(e)-Al-lY)”-“Something)”-“(College)”-Graduate)”-∫to the simple truth because∫-“(While-“(Standing)”-Near)”-“The-“(Industrial-“(Technology)”-Building)”-On)”-“The-“(University)”-Of-“(Northern-“(Iowa)”-Campus)”-In-“(1979),” “(I)”-∫sensed at audible levels∫-“(U)”-“(Are-“(R(e)-Al-lY”-“(Messed)”-“Up).” “(Although-“(I)”-“(Did”-“(Not)”-

“Deny)”-“(I)”-“(Was-“(A)”n-“(e)M-otional Intellectual)”-Mess),” “(My)”-“(sosO)”-“(R(e)-sponse)”-To)”-∫the simple truth of god was∫-“(To-“(Say)”-To-“(Myself),” “(sO)”-What),” “(sO)”-Is-“(everybody)”-else).”

the significance of∫-“(My)”-“(Lack)”-Of)”-“(Con)”-cern)”-“(Lack)”-Of)”-“(Interest)”-“(sO”-“(What)”-“(R(e)-sponse)”-∫is integrated into∫-“(sosO)”-∫becoming∫-“(5050)”-∫with a simple change that transforms the visual appearance of the letter s into the numeral five because∫-“(I)”-“(Was-“(Not-“(R(e)-ady)”-∫to undo∫-“The-“(Severe-“(eI)”-Unbalanced)”-“(Con)”-dition)”-Within)”-∫caused by∫-“The-“(Lack)”-Of)”-∫lamal by making the most difficult journey∫, ∫the inward journey∫. the creator of∫humanity∫asks∫-“The-“(e)X-ternally)”-Passioned)”-“(egO)”-tistical)”-“(Cold)”-“(egO”-“(e)M-otionless)”-“(Proud)”-“(I)”-“(Cons)”-∫of∫-“(Untrue)”-Blue)”-Humanity)”-∫to try to∫understand∫, ∫the inner beauty of being the internally passioned∫trusting∫tender∫truthful∫true blue child at play in the field of lavender hue∫was integrated into∫the gene code of humanity∫because∫the inner beauty∫allows∫each∫every individual in the we of all humanity to find genuine happiness∫joy∫fulfillment within∫.

the simple integrated truth includes the yellow color coding for the 38th element strontium on∫-“The-“(Periodic)”-Table)”-Of-“(e)L-ements)”-∫because∫-“The-“(Ye-ll-oW-“(Color)”-Coding)”-∫means∫-“The-“(He)”-Art)”-“(Con)”-Artists)”-∫are∫-“(Frozen)”-In)”-“(the-(Yellowish-Fear)-of-not-feeling-worthy-of-being)”-∫lovnewed∫. similar to strontium being used to produce a brilliant red color in fireworks, “The-“(Severely-“(eI)”-Unbalanced)”-“(I)”-“(Cons)”-“(Con)”-pensate)”-For-“(Being)”-“(Frozen)”-In)”-“(the-(Yellowish-Fear)”-“(Con)”-sumed)”-With)”-“(Seeing-Red-Angry)”-“(By-“(Con)”-trolling)”-“(the-(Orange-(emotional)-Flame)-Fire-Breathing-Dragon)”-“(Well-“(e)N-ough)”-To-“(Present)”-“(A)”-“(Brilliantly”-“(Choreographed)”-“Display)”-Of)”-“(e)X-ternal)”-Passion).”

the word trust being composed of the capital letter∫-“(T)”-∫followed by the word rust, “(T)”-Rust),” is not coincidental because the creator is trying to help∫the we of all humanity understand∫, similar to the reddish brown color of rust being caused by the reaction of iron materials to oxygen∫moisture in the air, ∫the child at play in the field of lavender hue deep within me at birth∫-“(R(e)-Acts)”-To)”-“(the-black-emotional-hole-threat)”-∫by∫-“(Becoming)”-“(A)”-“(Sh-It/Un-It-“(egO)”-God)”-“(Holding)”-“The-“(Iron”-“(Cross)”-“Position)”-For)”-“The-“(Boy”-“God”-“Con”-“Ta/Al).” the simple integrated truth is∫-“The-“(Con)”-Curious)”-“(Boy”-“God”-“Con”-“Ta/Al)”-Self)”-Preservation)”-Process)”-Of-“(Intellectually-“(Con)”-verting)”-∫true blue compassion∫courage∫into∫-“(Con)”-Passionate)”-“(Con)”-Courageous)”-Iron)”-Maidens)”-∫has unintended∫-“(Con)”-sequences)”-∫because the chemical symbol O for oxygen being the middle letter in god, moisture being analogous to fluidity means instead of∫living a fluid life based on∫sensing the whisper of god, “The-“(Severely-“(e)M-otionally Intellectually)”-Unbalanced)”-

“(Proud)”-“(I)”-“(Con)”-“(Intellectually)”-“Overides)”-∫the whisper by∫-“(Ridgedly)”-“Trying)”-To-“(Con)”-trol)”-“(the-emotional-deficit)”-∫with∫-“(Intellectually)”-earned)”-“Affection”-“Approval”-“Attention).”

the simple integrated truth in the word tender beginning with ten∫tend∫end, becoming the capital letter∫-“(T)”-∫followed by the word end, “(T)”-“(e)nd),” the interjection∫-“(e)R)”-∫is ∫the emotionally∫intellectually balanced child at birth∫has∫-“(A)”-“(Tenish”-“(Intense)”-“Tendency)”-To-“(Try)”-To-“(end)”-“(feeling-like-“(a)”-worthless-useless-unwanted-unneeded-unloved-nothing)”-∫with∫-“(This-“(e)R)”-That)”-“(Boy”-“God”-“Con”-“Ta/Al)”-“(e)X-ternal)”-Fix)”-∫because∫-“The-“(Timid-Tim-Mr.-“(T)”-Titans)”-∫were∫-“(Taught)”-“(Parented)”-“(Trained)”-∫to∫-“(Understand)”-“The-“(First/Only-“(Boy”-“God”-“Con”-“Ta/Al)”-“(He)”-Art-Law)”-Of-“(Thou”-“(Shalt)”-“earn)”-“Affection”-“Approval”-“Attention)”-To-“(Feel”-“(Worthy)”-Of-“(Being)”-∫lovnewed∫“(Must”-“(Be)”-“Fulfilled)”-“A”t-“(Any)”-“Cost).” the creator of∫humanity∫asks∫each∫every individual in the we of all humanity to try to understand∫the sense of∫-“(emotional-unbalance-within)”-∫is in an infinitesimal way caused by the vibrational attributes of the elements associated with∫the human quality of∫being tender.

the simple truth integrated into tenderness is the suffix ness including the element symbol, se, for selenium backwards was integrated to help∫humanity understand the individual being me at birth∫not touched with∫lamal∫will start∫-“(Getting)”-∫being me∫-“(Backwards)”-“(before-th(e)-age-of-two)”-∫because∫being me∫is dependent upon∫the nature of the child at play in the field of lavender hue at birth being nurtured with lamal∫. the green color coding of the element selenium on the peridodic table means∫the emotionally∫intellectually balanced child at birth∫was∫-“(e)N-couraged)”-∫to∫-“(Use)”-“The-“(Green)”-“To-“(Quench)”-“(the-(Orange-(emotional)-Flame)”-∫of∫-“(Yellowish-FearSeeing-Red-Anger)”-∫because∫-“(A”s”-“(Long)”-“A”s)”-“The-“(Severely-“(eI)”-Unbalanced)”-“(Proud)”-“(I)”-“(Cons)”-“(Do-“(Not”-“(Outwardly)”-“Appear)”-To-“(Be)”-Too-“(e)N-chanted)”-“(e)N-thusiastic)”-“(e)N-namored)”-With)”-“The-“(Green),” “(Appropriately”-“(e)N-gaging)”-In-“(Achieving”-“Attaining”-“Acquiring)”-“(More Not)”-Less)”-“(egO)”-tistical)”-Materialism)”-“(e)N-ables)”-“The-“(Boy”-“God”-“Con”-“Ta/Al)”-∫to∫-“(Con)”-ceal)”-Being)”-“(A)”n-“(e)M-otionalIntellectual)”-Mess).”

the simple truth is the blue color coded element tin, sn, backwards in the tenderness suffix ness was integrated to define being∫-“(Cold)”-“(egO”-“(e)M-otionless)”-∫because∫-“The-“(Generational”-“(e)F-fect)”-“Of)”-“(Bl-“(A)”-cKWh-“(I)”-T(e)”-“(Parenting)”-“(With-“(Fear),” “(A)”-“(Lack)”-Of)”-∫lamal∫-“(before-th(e)-age-of-two)”-∫includes∫-“(Con)”-cealing)”-∫being∫-“(Frozen)”-∫in∫-“(the-(Yellowish-Fear)-of-not-feeling-worthy-of-being)”-∫lovnewed∫on∫-“The-“(Bl-“(A)”-cKWh-“(I)”-T(e)”-Inside)”-∫by being∫-“(e)S-

pecially)"-Cool)"-"(Calm)"-"(Collected)"-"(N-Ice)"-ʃonʃ-"The-"(Bl-"(A)"-cK
Wh-"(I)"-T(e)"-Outside)." the atomic number 50 for tin on the periodic table means
ʃthe child at play in the field of lavender hue at birthʃbecomesʃ-"(5050)"-Within)"-
ʃbecause instead ofʃlooking in the mirrorʃ-"(Bl-"(A)"-cKWh-"(I)"-T(e)"-Saint)"-Ill)"-
"(Parents)"-"(Have)"-For-"(Generations)"-Tried)"-To-"(Prove)"-"(I)"-"(A"m-"(Not)"-
"(A)"-"(worthless-useless-unwanted-unneeded-unloved-nobody)"-ʃbyʃ-"(Making)"-"The-
"(Mad)"-"(Angry-"(Fear)"-Filled)"-Dash)"-To-"(earn)"-enough"-"(Well)"-"Deserved)"-
"Affection"-"Approval"-"Attention)."

the lower case i in tin betweenʃ-"The-"(Bookend"-"(Capital)"-"Letters)"-"(TN)"-
ʃbecomingʃ-"(TeeN)"-ʃphonetically was integrated to meanʃ-"(Being)"-Able)"-To-
"(Deny)"-"(I)"-"(A"m)"-"(Severely-"(e)M-otionallyIntellectually)"-Unbalanced)"-Left)"-
"The (Boy"-"God"-"Con"-"Ta/Al)"-With)"-"The-"(Difficult)"-Task)"-Of-"(Managing)"-
"(the-(Orange-(emotional)-Flame)"-ʃofʃ-"(Yellowish-FearSeeing-Red-Anger)"-"(By-
"(Intellectually)"-"Presenting)"-"(A)"-"(Perfect-Pearl-"(He)"-Art-Act)"-That-Was-
"(Tenish"-"(Intense)"-"Perfect/Flawless)"-"(e)N-ough)"-To-"(Both-"(Con)"-ceal)"-"The-
"(I)"-(Is-"(Turned)"-On)"-Without-"(Outwardly)"-"Appearing)"-To-"(Be)"-"(A)"-
"(Troubled)"-Teen)"-On)"-"The-"(Brink)"-Of)"-"(A)"-"(Nervous)"-Breakdown)." "The-
"(Boy"-"God"-"Con"-"Ta/Al)"-Runs)"-With)"-"The-"(Idea)"-That)"-"(I)"-"(Can"-
"(Soldier)"-"On)"-By-"(Keeping)"-"(the-(Orange-(emotional)-Flame)"-Volcano)"-From-
erupting)"-"(With-"(Membership)"-In)"-"(This-"(e)R)"-That)"-"(Mc-W(e)Mc-uS)"-"(I)"-
Pod)"-ʃbecauseʃ-"(Becoming)"-"The-"(egO)"-tistical)"-"Somebody)"-In-"(This-"(e)R)"-
That)"-"(Mc-W(e)Mc-uS)"-"(I)"-Pod)"-"(Pro-Wry-Dr.-"(I)"-Provides)"-"(A)"-
"(Heightened"-"(False)"-"Sense)"-Of-"(e)M-otional)"-Security)."

the simple truth isʃ-"(Becoming)"-"(A)"-"(Proud)"-"(P)"-ʃinʃ-"(This-"(e)R)"-That)"-"(Mc-
W(e)Mc-uS)"-"(I)"-Pod)"-To)"-"(feel-less-worthless-useless-unwanted-unneeded-
unloved)"-"(Has-"(A)"n-"(Unwanted"-"(Side)"-"(e)F-fect)"-ʃbecauseʃ-"(Pretending)"-
everything)"-Is)"-"(A)"-"(O.K.)"-"(Just)"-Fine)"-"(A"s"-"(Long)"-"A"s)-"(I)"-"(A"m)"-
"(e)N-thusiastic)"-"(e)N-dearing)"-"(e)N-gaging-"(e)N-ough)"-To-"(Get)"-"(U)"-"(e)N-
amored-"(e)N-ough)"-To-"(Dispense"-"(e)N-ough)"-"Affection"-"Approval"-"Attention)"
-ʃmeansʃ-"The-"(Severe-"(eI)"-Divide)"-Within)"-ʃwill not change one iota. the
metaphorʃ-"(e)M-pty)"-"A"s-"(A"n-"(Old"-"(Tin)"-"Can)"-ʃis meaningful becauseʃ-"The-
"(egO"-"(e)M-ptiness)"-"(Unhappiness)"-"(Lonliness)"-"(deep-sadness)"-ʃassociated
withʃ-"The-"(Generational"-"(e)F-fect)"-"Of)"-"(Bl-"(A)"-cKWh-"(I)"-T(e)"-
"(Parenting)"-"(With-"(Fear)," "(A)"-"(Lack)"-Of)"-ʃlamalʃ-"(before-th(e)-age-of-two)"-
ʃwill remain unchanged.

the simple integrated truth is the word truth being composed of the capital letterʃ-"(T)"-ʃfollowed by the word ruth, "(T)"-Ruth)," is not coincidental because although the creator tries to guideʃthe individuals in the we of all humanityʃto the simple truth with the whisper of the me, "(Holding)"-"The-"(Iron"-"(Cross)"-"Position)"-For)"-"The-"(Boy"-"God"-"Con"-"Ta/Al)"-Transforms)"-ʃthe child at play in the field of lavender hue deep within me at birthʃinto anʃ-"(egO)"-tistical)"-"Drunkard)"-"(Filled"-"(With)"-"Ruth)," defined asʃ-"(sorrow-grief-sadness)"-"(Self)"-Pity)"-"(R(e)-morse)"-"(Con)"-trition)." the integration of ru into truth, the chemical symbol for number 44 on the periodic table of elements, ruthenium, becoming ru-then-ium is intended to helpʃhumanity askʃanswer the simple question are you being meʃbecauseʃthe child at play in the field of lavender hue deep within me at birth understandsʃ, "(If)"-"(I)"-"(A"m-"(Not)"-ʃtouched with genuine true blue compassionʃ, thenʃ-"(I)"-"(Will)"-Be)"-"(Too-"(Gray)"-Zone)"-"(Con)"-fused)"-"(Con)"-flicted)"-"(Con)"-founded)"-"(To-"(I)"-"(uM)"-Know)"-"(Who-"(I)"-"A"m)"-"(Where-"(I)"-"A"m-To-gO)"-"(What-"(I)"-"A"m-To-Do)."

the god factor table above lays bare nine different factors associated with the word god, which means two numbers multiplied by each other have a product of 684. the factors of note on the table are 9ʃ76, 3ʃ228 because 76ʃ228 are both one more than the values of 75ʃ227 for whisperʃsense found by summing the atomic numbers associated with the element symbols in the words whisperʃsense on page 31ʃ32 respectively.

the god factor

684	1	684
684	2	342
684	3	228
684	4	171
684	6	114
684	9	76
684	12	57
684	18	38
684	19	36

the simple integrated truth is the factor 57 on the tableʃis meaningful to the authorʃbecause Lanthanum, which is atomic number 57 on the periodic table with the chemical symbol La, beginning with Lan, an acronym in the computer industry for local area network, is not coincidental. the simple truth is the termʃ-**'THE-'{(LAN'-'NET'-'WEB)}'-**ʃis used throughout the book onʃ-**'{(MIND)}'-'{(CON)}'-TROL)}'-**ʃtitledʃ-**'{(S-LAUGHTER)}'-**ʃto helpʃ-"(Humanity)"-ʃunderstand the special la di da at birthʃchose withʃthe blessing of free willʃto becomeʃ-‘THE-‘{(MENTAL)}’-‘{(ME)}’-‘{(TAL)}’-‘{(CON)}’-PUTER)}’-ON-‘THE-‘{(MISSION)}’-OF-‘{(S-LAUGHTERING)}’-“(U)”-“(Simple)”-Minded)”-“(Mental)”-Midgets)”-ʃforʃ-‘{(FUN’-‘SPORT’-‘{(E)}N-TERTAINMENT)}’-WITH-‘THE-‘{(WEAPON)}’-OF-‘{(MIND)}’-‘{(CON)}’-TROL)}’-ʃbecause ofʃ-“The-“(Lack)”-Of)”-ʃlamalʃdefined asʃgenuine compassionate emotionalʃphysical touchʃassociated withʃ-“The-“(Generational”-“(e)F-fect)”-“Of)”-“(Bl-“(A)”-cKWh-“(I)”-T(e)”-“(Parenting)”-“(With-“(Fear),” “(A)”-“(Lack)”-Of)”-ʃlamalʃ-“(before-th(e)-age-of-two).”

althoughʃ-“The-“(Author)”-Has-“(Begged/Pleaded)”-ʃgod to endʃ-‘THE-‘{(MIND)}’-‘{(CON)}’-TROL)}’-TORTURE)}’-INFLICTED)}’-UPON)}’-“The-“(Author)”-For-“(Decades),” which includesʃ-“(Sleep)”-Deprivation)”-“(e)M-otionalPhysical)”-Pain),” ʃthe author now understandsʃthe resonant frequency of factor 57 plays a crucial role in ʃwhyʃthe creator can do nothing to endʃ-‘THE-‘{(HORRENDOUS)}’-TORTURE)}’-OF-‘THE-‘{(MIND)}’-‘{(CON)}’-TROLLING)}’-PIED)}’-PIPER)}.’ the simple truth is the only solution to endʃ-‘THE-‘{(DEPRAVED)}’-ACTIONS)}’-OF-‘THE-‘{(SACRIFICIAL)}’-LAMB)}’-WITH-‘THE-‘{(MESSIAH-‘{(CON)}’-PLEX)}’-ʃis to ʃchoose with the blessing of free will to removeʃ-“The-“(Boy”-“God”-“Con”-“Ta/Al)”-“(egO”-“(Bridge)”-“Fix)”-ʃforʃ-“(the-black-(emotional)-hole-problem-of-(Yellowish-FearSeeing-Red-Anger)-deep-within)”-ʃbecauseʃthe author understandsʃnothing will change withinʃ-“(the-feeling-like-“(a)”-nothing)”-“(egO)”-tistical)”-“Somebody)”-ʃuntilʃ-“The-“(Proud)”-“(I)”-“(Con)”-ʃfirstʃacceptsʃacknowledgesʃadmitsʃ-“(I)”-ʃamʃ-“(Severely-“(e)M-otionallyIntellectually)”-Unbalanced).”

ʃthe author finally understoodʃthe significance of the car radio cutting out forʃ-“(nO-“(Ap-“(Parent)”-Reason)”-ʃin 1998 becauseʃtrue blue compassionʃbeing defined byʃlaʃ, the wordʃ-“(LacK)”-ʃin black spelledʃ-“(B-La-cK)”-ʃbeginning withʃ-“(La),” “(La)”-ʃbeing the chemical symbol for number 57 lanthanum on the periodic table of elements is not coincidental. the simple integrated truth isʃ-“The-“(nO-“(Ap-“(Parent)”-Reason)”-ʃisʃ-“The-“(Generational”-“(e)F-fect)”-“Of)”-“(Bl-“(A)”-cKWh-“(I)”-T(e)”-“(Parenting)”-“(With-“(Fear),” “(A)”-“(Lack)”-Of)”-ʃlamalʃ-“(before-th(e)-age-of-two)”-ʃbecauseʃ-‘THE-‘{(CLOCKWORK)}’-ORANGE)}’-‘{(MIND)}’-‘{(CON)}’-TROLLING)}’-PIED)}’-PIPERS)}’-PREY)}’-UPON)}’-“The-“(Boy”-“God”-“Con”-“Ta/Al)”-“(egO”-“(Bridge)”-“Fix)”-ʃforʃ-“(the-black-(emotional)-hole-problem-of-(Yellowish-FearSeeing-Red-Anger)-deep-within)”-ʃcaused byʃ-“The-“(Lack)”-Of)”-ʃlamalʃdefined asʃgenuine compassionate emotionalʃphysical touchʃ.

the simple truth integrated intoʃhuman languageʃincludes the word path integrated into telepathic, which defines the ability ofʃhuman beingsʃto both sense the whisper of god ʃcommunicate telepathically with vibrations receivedʃtransmitted through the skullʃ. ʃthe author has no doubtʃtelepathic communication exists becauseʃ-‘THE-

‘{(CLOCKWORK)}’-ORANGE)}’-‘{(MIND)}’-‘{(CON)}’-TROLLING)}’-PIED)}’-PIPERS)}’-ARE-‘{(DEMENTED’-‘DELUSIONAL’-‘DERANGED)}’-ENOUGH)}’-TO-‘{(BOY’-‘GOD’-‘CON’-‘TA\AL)}’-RATIONALIZE)}’-USING-‘{(MENTAL)}’-TELEPATHY)}’-∫to∫-‘{(CON)}’-MIT)}’-‘THE-‘{(PERFECT)}’-CRIME)}’-BY-‘{(KILLING)}’-“(U)”-“(Obstinate)”-Oafs)”-‘{(WITHOUT’-‘{(EVEN)}’-‘BEING)}’-‘A’T-‘THE-‘{(SCENE)}’-OF-‘THE-‘{(CRIME)}’-PROVES-‘{(I)}’-‘A’M-‘{(GOD)}.’ ‘THE-‘{(SUBTLETY)}’-OF-‘{(MIND)}’-‘{(CON)}’-TROL)}’-∫is such that distinguishing between∫-‘{(MIND)}’-‘{(CON)}’-TROL)}’-PLANTED)}’-THOUGHTS)},’ sensing the whisper of the one true god∫the me, thoughts of internal origin is extremely difficult because∫-‘THE-‘{(LAN’-‘NET’-‘WEB)}’-‘{(PRO-WRY-‘{(DR.-JEKYLLSMR.-HYDES)}’-HAVE-‘{(BOY’-‘GOD’-‘CON’-‘TA\AL)}’-‘PERFECTED)}’-BEING-ABLE-TO-‘{(MENTALLY)}’-LINK)}’-TO)}’-∫the sensory inputs, including visual∫auditory∫tactile inputs of∫-“(U)”-“(Worthless”-“Useless”-“Witless)”-Twits)”-‘{(ALMOST-‘{(INSTANTAINEOUSLY)},’ ‘{(IMMEDIATELY)}’-AFTER)}’-∫cognitive recognition∫mental processing of the inputs has begun.

> the one true god∫the me of the trust∫tenderness∫truth asks∫-“(Untrue)”-Blue)”-Humanity)”-∫to try to∫-“(Bl-“(A)”-cKWh-“(I)”-T(e)”-Saint)”-Ill)”-“(Understand),” ‘THE-‘{(SUBTLETY)}’-OF-‘{(MIND)}’-‘{(CON)}’-TROL)}’-∫must never be underestimated because distinguishing between∫-‘{(THOUGHTS)}’-INDUCED)}’-BY-‘THE-‘{(MIND)}’-‘{(CON)}’-TROLLING)}’-PIED)}’-PIPERS)},’ sensing the whisper of the one true god∫the me, self-originated thought is virtually impossible without first becoming∫emotionally∫intellectually balanced within∫.

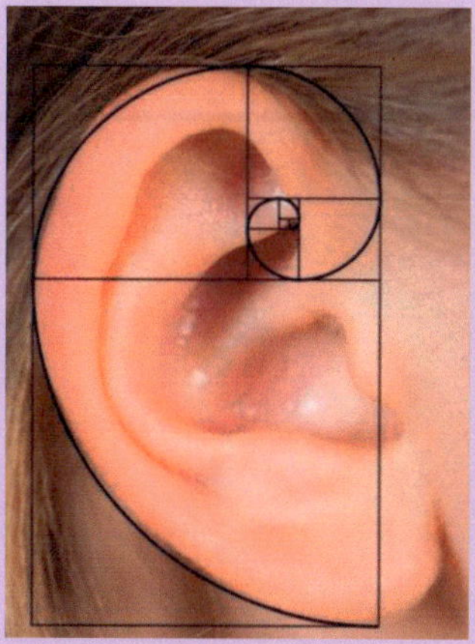

the simple truth integrated into∫human language∫includes the word path integrated into the auditory pathway because one function of∫human hearing∫is to provide a pathway to sense the trusting∫tender∫truthful whisper of the one true god∫the me. the simple truth is the fibonacci spiral of the human ear, as shown in the image above, is an integral part of sensing the whisper of the me because the shape of the human ear is proportional to the fibonacci spiral of the chambered nautilus shell used to define the energy flow within god.

the spiraling shape of the cochlea, resembling a snail shell, in the inner ear was also created to sense the trustingʃtenderʃtruthful whisper of the one true godʃthe me because vibrationally sensitive hair cells in the cochlea change the vibrations of sound waves into the nerve impulses sent to the vibrationally sensitive cells in the brain. researchers have determined the cochlear spiral produces a "whispering gallery mode," which is similar to quiet sounds traveling longer distances along a curved wall in a building without losing energy. therefore a primary function of the spiraling shape of the cochlea is to increase sensitivity to low frequencies, such as a 1hz to 10 hertz whisper of god, by as much as 20 decibels.

although the frequency range of audible human hearing is generally accepted to be between 20 to 20,000 Hz, research found the sensitivity to vibration is much broader than the audible range of human hearing because sensitivity to frequencies as low as 1.5 Hz have been measured. the distinction between hearing audible soundʃsensing the whisper of god is sensing is done at a frequency below the 20 Hz threshold of audible sound. in addition to finding humans have a whole body vibration of 5-10 hertz, research shows 10 hertz is the frequency of the human brain in a calm state.

the simple truth integrated intoʃhuman languageʃincludes the word path integrated into the sympatheticʃparasympathetic nervous system. the nervous system ofʃhuman beingsʃis composed of the central nervous system, the peripheral nervous system. while the central nervous system is composed of the brainʃthe spinal cord, the peripheral nervous system is subdivided into the somaticʃautonomic systems. the basic distinction between the somaticʃautonomic systems is the somatic system guides voluntary muscularʃsensual movements, such as soundʃsmellʃtasteʃtouch, the autonomic system regulates involuntary movements, such as breathingʃheart rateʃdigestion.

since the sympatheticʃparasympathetic systems are autonomic, the systems affect the bodily functions that keepʃhumans aliveʃautomatically without anyʃintellectual inputʃ. therefore the creator can use the systems to help guideʃall of the individuals in the we of all humanity from withinʃwith the trustingʃtenderʃtruthful whisper of the creator because ʃhumansʃdo not have the ability toʃ-"(Intellectually)"-"(Con)"-trol"-ʃthe sympatheticʃ parasympathetic nervous systems. the scientific term sympathetic vibration, also called resonance, defines the interrelationship between the whisper of godʃthe cells in the trustingʃ tenderʃtruthful path because the whisperʃthe cells in the path must have the same natural frequency or a vibration that is a multiple of the frequency.

although the autonomic nervous system is not associated with either the sense of hearingʃ telepathic communication, sensory input is received through the skin because autonomic nerves affect the muscles that cause hair to stand up, commonly called "goose bumps," as well as the ability of the body to sweat. similar to the tympanic membrane in the ear, the skin on the body is analogous to the tight skin on a drum because being awash in the whisper wave, which in simple terms is a tender hug from the creator, counteractsʃ-"The-"(Boy"-"God"-"Con"-"Ta/Al)"-"(egO"-"(Bridge)"-"Fix)"-ʃused toʃ-"(Wall)"-Out)"-"(the-black-emotional-hole-problem)"-ʃbyʃhaving a calming effect on human physiologyʃ. the interrelationship between the whisper waveʃthe sympathetic calming effect at the cellular level helps explainʃthe calm mental stateʃ-"(I)"-ʃexperienced onʃ-"The-"(Campus)"-Of-"(Iowa"-"(State)"-"University)"-ʃin 1967.

the simple truth isʃthe calm mental state needed to senseʃthe tender whisper of the one true godʃthe me of the trustʃtendernessʃtruthʃis documented in the book titledʃthe simple integrated truth becauseʃthe author gives several examples of sensingʃthe whisperʃwhile calm mentallyʃ. ʃthe first example given occurs in 1967 while walking onʃ-"The-"(Campus)"-Of-"(Iowa"-"(State)"-"University)"-ʃbecauseʃ-"(I)"-ʃsensedʃyou should be an architect. "(My-"(R(e)-sponse)"-Was-"(I)"-"(Do"-"(Not)"-"Want)"-To-"(Spend"-"(5)"-"Years)"-In-"(College-"(Getting)"-"(A)"-"(Degree)"-In-"(Architecture)"-ʃbecauseʃinstead of guiding me to understand identifyingʃdevelopingʃusing my individual internal passion in architecture was in the best interest of meʃthe meʃhumanityʃ, "(I)"-"(Was)"-"(Parented)"-"(To"-"(Be)"-"Too)"-"(Gray)"-Zone)"-"(Con)"-fused)"-"(Con)"-flicted)"-"(Con)"-founded)"-"(To-"(Know)"-"(Who-"(I)"-"A"m)"-"(Where-"(I)"-"A"m-To-gO)"-"(What-"(I)"-"A"m-To-Do)."

the table on the following page applies the atomic numberʃelement symbol on the periodic table of elements for the words nervesʃnervous. the 114 sum for the word nerves, which applies to nerves in general, is meaningful because 114 is one of the god factors in the table on page 25. the asterick associated with vanadium in the word nerves means the phonetic ve, is used in place of the chemical symbol v for vanadium. the sum of 75 for nitrogen ʃerbium in both nervesʃnervous being the same as the total for the word sense on page 32 is not coincidental because nerve cells must be sensitive to the whisper of god.

the interrelationship between the trustingʃtenderʃtruthful path of the creator ofʃhumanityʃ, the centralʃperipheral nervous system is evidenced by the o in god defined by atomic number 8 because o is also present in the word nervous. the sum of the atomic numbers for erbiumʃsulfur, 68ʃ16, being 84 further substantiates the interrelationship because 84 is the last two digits in the number 684 derived by substituting the atomic numbers for gadoliniumʃoxygen, gdʃo, used to define the word god.

the first∫last letter of nervous, n∫s, being the same as the center letter in sense∫whisper respectively is not coincidental because each of the three trusting∫tender∫truthful pathways is associated with the central∫peripheral nervous system. the simple integrated truth is the er in nervous is also not coincidental because the word whisper ending with er helps establish the interrelationship between the creator of∫humanity∫, the trusing∫tender∫truthful path at the cellular level. the sum for nervous, 214, being 100 more than the sum for nerves, 114, is associaed with the atomic numbers for uranium∫oxygen being 92∫8 respectively.

although uranium has the highest atomic number, 92, in the word nervous, the symbol u for uranium is not present in the words previously defined. the significance of uranium being the only radioactive element in nervous is the simple truth associated with the car radio of the author cutting out for∫-"(nO-"(Ap-"(Parent)"-Reason)." the other interrelationship of significance is the color coding for uranium on the periodic table because the∫lavender∫ color code for uranium also defines∫the child at play in the internally passioned∫trusting∫ tender∫truthful∫true blue humble∫compassionate∫curious∫courageous∫zestful field of lavender hue within∫.

nerves∫nervous

word	element	symbol	number	color
nerves				
	nitrogen	n	7	green
	erbium	er	68	pink
	vanadium	ve*	23	red
	sulfur	s	16	green
total			114	
nervous				
	nitrogen	n	7	green
	erbium	er	68	pink
	vanadium	v	23	red
	oxygen	o	8	green
	uranium	u	92	lavender
	sulfur	s	16	green
total			214	

after the same chemical symbol∫atomic number methodology is used to define the whisper of the one true god∫the me, the sum of the six element symbols for tungsten, w, hydrogen, h, iodine, i, sulfur, s, phosphorus, p, erbium, er, in whisper spelled whi-s-per, is 227. although scientists have determined that resonance occurs when two waves have the same natural frequency or a multiple of the vibration, when the number to sense the whisper of god, 75, shown on page 32 is multiplied by 3 the result is 225, which is 2 from the elemental total for the whisper of god, 227.

the word whisper beginning with the same two letters, wh, as why, is not coincidental because the total for the wh elements in whisper, 75, being the same as the sum of the elements used to define sense is interpreted to mean the one hertz message frequency is included in 227 because in addition to nitrogen being number 7 on the periodic table, the sum of 3+4 in 34∫7∫34 is 7. the seven in 227∫75, 34∫7∫34 means nitrogen is a critical element in sensing the whisper of god. alternately if 1 hertz for the whisper of god is added to 75∫227, the numbers for whisper∫sense become 228∫76 respectively, which are both factors of 684 in the table on page 25.

the whisper

word	element	symbol	number		color
whisper					
whi	tungsten	w	74		red
	hydrogen	h	1		white
	iodine	i	53		green
	subtotal			128	
s	sulfur	s	16		green
	subtotal			16	
per	phosphorus	p	15		green
	erbium	er	68		pink
	subtotal			83	
total				227	

the simple truth integrated into∫the ability of human beings to sense∫the whisper of god using the chemical symbol∫atomic number methodology is shown in the table for sense on the following page. the atomic numbers associated with the two chemical symbols in se-n-se, selenium, se, nitrogen, n, are 34∫7 respectively. the sum of the atomic numbers for each of the element symbols comprising the word sense, 34∫7∫34, is 75. the sum of 34 plus

34 in sense equaling 68, which is both the atomic number for erbium in whisper, the first two digits, 68, in 684 is not coincidental.

sense

word	element	symbol	number	color
sense				
	selenium	se	34	green
	nitrogen	n	7	green
	selenium	se	34	green
total			75	

as shown in the table below, the simple integrated truth is the elemental composition of arsenic∫potassium used to create a cell capable of sensing the trusting∫tender∫truthful whisper of the one true god∫the me is integrated into the word ask because the integration of∫free will∫into∫the human gene code∫means the creator must ask∫cannot make any∫human being do that which is in the best interest of me∫the me∫the we of all humanity∫. the simple truth integrated into the green∫orange color coding for arsenic∫potassium respectively is the elemental composition plays an infinitesimal part in∫-“The-“(Boy”-“God”-“Con”-“Ta/Al)”-Using)”-“The-“(Green)”-∫to∫-“(Sedate)”-“(the-(Orange-(emotional)-Flame)-within)”-∫because creating cells sensitive to the whisper of the one true god∫the me is dependent upon the vibrational attributes of each element.

ask

word	element	symbol	number	color
ask				
	arsenic	as	33	green
	potassium	k	19	orange
total			52	

the simple truth integrated into the periodic table includes every letter in the word poison being the symbol for a green color coded element because the letters∫-“(P-O-I-S-O-N)”-∫are the symbols for number 15 phosphorus, number 8 oxygen, number 53 iodine, number 16 sulfur, number 8 oxygen, number 7 nitrogen, which sum to 107. the creator asks∫the we of all humanity to try to understand∫allowing∫-“The-“(Boy”-“God”-“Con”-“Ta/Al)”-To-“(Use)”-“The-“(Green)”-∫to∫-“(Prevent)”-“(U)”-“(From-“(Seeing)”-“(I)”-“(A”m)”-“(A)”-“(Severely-“(eI)”-Unbalanced)”-“(Orange-(emotional)-Flame)-Driven)”-Mess)”-∫is not without∫-“(Con)”-sequences)”-∫because∫-“The-“(Green)”-“(M-One-Y)”-∫for∫-“The-“(egO)”-tistical)”-One)”-“(Poisons)”-“The-“(Chances)”-∫of being guided to∫find genuine happinessʃjoyʃfulfillment withinʃby sensing the whisper of god. the sum of 52 for the word ask being equal to the green color coded element tellurium, te, used to define the qualities of trustʃtendernessʃtruth is not coincidental.

the table below shows the simple truth integrated intoʃhuman languageʃincludes the word why, pronounced '(h)wī phonetically, being defined by the chemical symbolsʃatomic numbers on the periodic table of elements. the elemental sum of 114 for number 74 tungsten, w, number 1 hydrogen, h, number 39 yttrium, y, that compose the word why, is meaningful because 114 is one of thc factors for 684 in the table on page 25 used to define the interrelationship between godʃthe individuals in the we of all humanityʃ. the one true godʃthe me asksʃthe individuals comprising the we of all humanityʃto beʃtrue blue curious by asking why humanityʃ-"(Suffers)"-ʃwithʃthe true blue purity ofʃintent because the creator will guideʃeachʃevery individual toʃthe simple truth.

why

word	element	symbol	number	color
why				
	tungsten	w	74	red
	hydrogen	h	1	white
	yttrium	y	39	red
total			114	

ʃthe child at play in the field of lavender hue deep within me at birthʃbecomingʃ-"The-"(egO)"-tistical)"-"One)"-ʃis integrated into the phonetic distinction between thəʃthē because while thə, spelled with the upside downʃbackwards schwa symbol, ə, is usually pronounced with the 'uh' sound, thē usually used before vowels pronounced with a long phonetic ē is used to suggest definitive uniqueness. the simple truth is although there is definitely only one the meʃone the we of all humanityʃwithin the universe of the one true godʃthe me of the trustʃtendernessʃtruth, the is pronounced thə because the humble one true godʃthe me did not integrateʃ-"(egO)"-ʃintoʃthe gene code of the individuals comprising the we of all humanityʃ.

the function word thē pronounced with the long phonetic ē used before vowels for emphasis is intended to helpʃ-"(Humanity)"-ʃunderstandʃ, ʃthe true blue individual me at birthʃ-"(Places"-"(e)M-phasis)"-"On)"-"Taking)"-"The-"(Boy"-"God"-"Con"-"Ta/Al)"-"(Vow)"-To-"(earn)"-"Affection"-"Approval"-"Attention)"-"(before-th(e)-age-of-two)"-ʃbecauseʃ-"The-"(egO"-"(Bridge)"-"Fix)"-Of)"-"The-"(Boy"-"God"-"Con"-"Ta/Al)"-Serves)"-"The-"(Function)"-Of-"(Attending)"-To)"-"(the-black-emotional-hole-problem-of-feeling-unwanted-unneeded-unloved).” the simple integrated truth is thē with the long phonetic ē defines∫-“The-“(Severely-“(e)M-otionallyIntellectually)”-Unbalanced)”-“(Proud)”-“(I)”-“(Cons)”-“(Who)”-Know)”-“(I)”-“(A”m-“(Definitely”-“(egO)”-tistically)”-“Superior)”-To)”-“(U)”-“(They”-“Them”-“Those)”-Others)”-∫because∫-“(egO)”-Becomes)”-“The-“(Wonder)”-Drug)”-∫that allows∫the true blue humble human being at birth∫to-“((Proudly)”-“Proclaim)”-“(I)”-“(A”m)”-“Somebody).”

the simple truth is the capital letter∫-“(T)”-∫in the word thē, spelled∫-“(T-“(He),” was integrated to mean∫the true blue compassionate∫curious∫courageous child at birth∫is∫-“(Holding)”-“The-“(Iron”-“(Cross)”-“Position),” the capital letter∫-“(T)”-∫in∫-“(T-“(He),” for∫-“The-“(Boy”-“God”-“Con”-“Ta/Al),” the word∫-“(He)”-∫in∫-“(T-“(He)”-∫because∫-“(After)”-“The-“(Boy”-“God”-“Con”-“Ta/Al)”-“(Con)”-verts)”-∫true blue curiosity∫into∫-“(Con)”-Curiosity),” “(He)”-“(Intellectually)”-“(Con)”-verts)”-∫true blue compassion∫courage∫into∫-“(Con)”-Passionate)”-“(Con)”-Courageous)”-“(He)”-Art)”-Act)”-“(Co-“(Con)”-spirators).” “The-“(Intellectual)”-“(Con)”-version)”-∫of∫true blue compassion∫curiosity∫courageous∫into∫-“(Con)”-Passion)”-“(Con)”-Curiosity)”-“(Con)”-Courage)”-“(R(e)-sults)”-∫in∫the true blue child at birth∫becoming∫-“(A)”-“(Cold)”-“(egO”-“(e)M-otionless)”-Human)”-∫not∫being me∫because∫-“(Holding)”-“The-“(Iron”-“(Cross)”-“Position)”-“(Intellectually)”-“(Traps)”-∫genuine emotion∫-“(Behind)”-“The-“(Closed)”-“(e)M-otional)”-“(Door)”-Deep)”-Within).”

the simple integrated truth in the word∫-“(He),” a masculine demonstrative pronoun∫definite article, beginning with the capital letter∫-“(H),” also being the chemical symbol for hydrogen, the only white element on the periodic table of elements is not∫-“(Bl-“(A)”-cK Wh-“(I)”-T(e)”-Coincidental)”-∫because∫-“(After)”-“(the-black-emotional-hole)”-“(Becomes)”-“(A)”-“(Serious”-“(enough)”-“Problem),” “The-“(Boy”-“God”-“Con”-“Ta/Al)”-“Knows)”-“The-“(H)”-Hour)”-For-“(Launching)”-“The-“(Specific”-“(Tactical)”-“Operation)”-Of-“(Installing)”-“The-“(egO)”-Bridge)”-Has-“(Arrived).” similar to hydrogen, the first element with the atomic number one, as well as the only white color coded element that does not belong to any one class, “The-“(Boy”-“God”-“Con”-“Ta/Al)”-Installation)”-Of)”-“The-“(egO)”-Bridge)”-“(R(e)-sults)”-In-“A”n-“(egO”-“(Inflated)”-“Sense)”-Of-“(Self),” integrated into the capital letter∫-“(I)”-∫in∫-“(Wh-“(I)”-T(e),” “(Which-“(Puts”-“(each/ every)”-“(Sh-It/Un-It-“(egO)”-God)”-“(In)”-“(A)”-“(Class)”-“(All)”-By-“(Myself)”-∫because∫the true blue humble individual me at birth∫becomes∫-“The-“(egO)”-tistical)”-One)”-On)”-“The-“(I)”-“(Robot)”-Mission)”-Of-“(Putting)”-Out)”-“(the-(Orange-(emotional)-Flame)”-“(With-“(Intellectually)”-earned)”-“Affection”-“Approval”-“Attention).”

when the same periodic table atomic number∫element symbol methodology used for the word god,684, is applied to the word thē, the chemical symbol for tellurium, tē, becomes the first∫last letter in thē, the chemical symbol for hydrogen, h, becomes the middle letter in thē. as shown in the table below, the atomic numbers for tellurium∫ hydrogen of 52∫1 respectively become 53 when added together. the simple integrated truth is the sum of 53 being the green color coded element Iodine with the capital letter I chemical symbol is not coincidental because∫-"The-"(Capital)"-Letter)"-"(I)"-∫defines∫-"The-"(e)X-aggerated)"-Sense)"-Of-"(Self)"-Importance)"-∫associated with∫-"The-"(Boy"-"God"-"Con"-"Ta/Al)"-"(egO"-"(Bridge)"-"Fix)"-∫for∫-"(the-black-emotional-hole-problem)."

"(Th(ē)"

Word	Element	Symbol	Number	Color
"(Th(ē)"				
	Tellurium	Te	52	Green
	hydrogen	h	1	white
Total	Iodine	"(I)"	53	Green

"The-"(Irate)"-"(I)"-"(Cons)"-∫of∫-"(Untrue)"-Blue)"-Humanity)"-∫with∫-"The-"(I)"-"(Do"-"(Too)"-"Rate)"-"Attitude)"-∫do not understand the true blue human being me at birth∫is∫-"(R(e)-duced)"-To-"(B(e)-Ing)"-"(A)"-"(Cold)"-"(egO"-"(e)M-otionless)"-∫true blue∫-"(Has)"-Been)"-"(before-th(e)-age-of-two)"-∫because∫-"The-"(Boy"-"God"-"Con"-"Ta/Al)"-"(egO"-"(Bridge)"-"Fix)"-For)"-"(the-black-emotional-hole-(Yellowish-Fear Seeing-Red-Anger)-problem)"-"(Makes-"(Ta/Al)"-"The-"(Lord/Master)"-Of)"-"(the-(Orange-(emotional)-Flame)." ∫the emotionally∫intellectually balanced human being at birth∫does not∫understand being human∫means∫-"The-"(egO"-"(Bridge)"-"Fix)"-∫of∫-"The-"(Boy"-"God"-"Con"-"Ta/Al)"-∫must be∫-"(Taken"-"(Out)"-"Of)"-"The-"(Loop)"-∫because∫instead of trying to understand∫-"The-"(Boy"-"God"-"Con"-"Ta/Al)"-∫is∫-"The-"(Problem)," "The-"(I)"-"(Robot"-"(Human)"-"Being)"-∫with∫-"The-"(I)"-"(Turned)"-On)"-∫is∫-"(Boy"-"God"-"Con"-"Ta/Al)"-"(Con)"-vinced)"-"(Ta/Al)"-Is)"-"The-"(Solution)"-To)"-"The-"(Problem)."

the capital letter∫-"(I)"-∫associated with the word thē pronounced with a long phonetic ē, lays bare being∫-"(egO)"-tistical)"-∫because the word∫-"(ēgO)"-∫beginning with the same long phonetic ē sound as thē is not coincidental. the simple integrated truth in the three capital letters of the english language from which the word∫-'[(**H**-"(A)"-**D**)}'-∫is comprised, define∫-"The-"(Internal-"(Con)"-struct)"-∫of each∫every∫-"(egO)"-tistical)"-"(I)"-"(Con)"-∫of∫-"The-"(Past"-"Present"-"Future)"-Population)"-∫on∫the home planet∫because the one true god∫the me is trying to help∫-"(Humanity)"-∫understand the we of all humanity∫has been∫-'[(**H**-"(A)"-**D**)}'-∫by∫-"The-"(Generational"-"(e)F-fect)"-"Of)"-"(Bl-"(A)"-cK

Wh-“(I)”-T(e)”-“(Parenting)”-“(With-“(Fear),” “(A)”-“(Lack)”-Of)”-ʃlamalʃ-“(before-th(e)-age-of-two).”

the simple truth is the distinction to be made betweenʃ-“The-“(Internal-“(egO)”-tistical)”-“(Con)”-struct)”-ʃdefined by the brown color coded capital letterʃ-“(A),” as compared toʃ-‘THE-‘[(INTERNAL-‘[(EGO)]’-TISTICAL)]’-‘[(CON)]’-STRUCT)]’-ʃdefined by the black color coded capital letterʃ-‘[(H)]’-ʃin aʃ-‘[(BOLD)]’-ʃfont, ‘THE-‘{(INTERNAL-‘{(CON)}’-STRUCT)}’-ʃdefined by theʃ-‘{(BOLD)}’-ʃwhite color coded capital letterʃ-‘{(D)}’-ʃinʃ-‘[(H-“(A)”-D)}’-ʃisʃthe blessing of free willʃ. whileʃ-“The-“(Internal-“(egO)”-tistical)”-“(Con)”-struct)”-ʃof all three are initially defined by the brown color coded capital letterʃ-“(A),” which is caused byʃ-“The-“(Generational”-“(e)F-fect)”-“Of)”-“(Bl-“(A)”-cKWh-“(I)”-T(e)”-“(Parenting)”-“(With-“(Fear),” “(A)”-“(Lack)”-Of)”-ʃlamalʃ-“(before-th(e)-age-of-two),” ‘THE-‘[(INTERNAL-‘[(EGO)]’-TISTICAL)]’-‘[(CON)]’-STRUCT)]’-ʃassociated with the capital letterʃ-‘[(H)],’ ‘THE-‘{(INTERNAL-‘[(EGO)]’-TISTICAL)]’-‘{(CON)}’-STRUCT)}’-ʃassociated with the capital letterʃ-‘{(D)}’-ʃinʃ-‘[(H-“(A)”-D)}’-ʃalso includesʃa choice made with the blessing of free willʃ.

the capital letterʃ-‘[(H)]’-ʃinʃ-‘[(H-“(A)”-D)}’-ʃdefinesʃ-‘THE-‘[(CARBON’-‘[(BL-‘[(A)]’-CK)]’-‘[(EGO)]’-TISTICAL)]’-‘[(INTENT)]’-ʃofʃ-‘[(BECOMING)]’-‘[(E)]X-TERNALLY)]’-PASSIONED)]’-ʃtoʃ-‘[(SCREW|FLEECE)]’-“(U)”-“(Insignificant)”-To)”-‘THE-‘[(I)]’-‘[(CON)]’-“(Others)”-ʃforʃ-‘[(SELFISH-‘[(EGO)]’-TISTICAL)]’-MATERIALISTIC)]’-GAIN)].’

the capital letterʃ-“(A)”-ʃinʃ-‘[(H-“(A)”-D)}’-ʃdefines the unintentionalʃ-“(egO)”-tism)”-ʃofʃ-“The-“(I)”-“(Con)”-ʃtoʃ-“(Intellectually”-“(earn)”-“enough)”-“Affection”-“Approval”-“Attention)”-To-“(Fill)”-“(the-black-emotional-hole).”

the capital letterʃ-‘{(D)}’-ʃinʃ-‘[(H-“(A)”-D)}’-ʃdefinesʃ-‘THE-‘{(EGO)}’-TISTICAL)}’-‘{(INTENT)}’-OF-‘THE-‘{(I)}’-‘{(CON)}’-ʃtoʃ-‘{(BECOME)}’-‘{(E)}X-TERNALLY)}’-PASSIONED)}’-TO-‘{(S-LAUGHTER)}’-“(U)”-“(Worthless”-“Useless”-“Vulgar”-“(A”n-I’m-“Als)”-ʃwithʃ-‘THE-‘{(WEAPON)}’-OF-‘{(MIND)}’-‘{(CON)}’-TROL)}’-ʃforʃ-‘{(FUN’-‘SPORT’-‘{(E)}N’-TERTAINMENT)}.’

the simple truth integrated intoʃhuman languageʃthat explains whyʃhumanityʃ-“(Suffers)”-ʃis laid bare by the similarity betweenʃlamalʃ, ʃthe word created to define genuine compassionate emotionalʃphysical touchʃ, the words ofʃjesus christʃfrom the cross, *əlahí əlahí ləmáh šəvaqtáni*, which are translated to meanʃ-“(My-“(God)”-“(My-“(God),” ʃwhyʃ-“(Have-“(You”-“(Forsaken)”-“(M(e)?”-ʃbecause the transliteration of *ləmáh*, meaning why is lama. the word lama becomingʃlamalʃby simply adding a lower case l explainsʃwhy jesus thoughtʃthe creator ofʃhumanityʃ-‘[(H-“(A)”-D)}’-“(Forsaken-“(M(e)”-ʃbecause instead of∫being touched with lamal∫,“(Jesus)”-∫was∫-“(Parented)”-“(With-“(Fear),” “(A)”-“(Lack)”-Of)”-∫lamal∫-“(before-th(e)-age-of-two)”-∫by∫-“(Bl-“(A)”-cKWh-“(I)”-T(e)”-Saint)” Ill)”-“(Parents).”

the simple integrated truth in *əlahi,*composed of *əlah* meaning god followed by *i* meaning my is the undeniable importance of∫touching the internally passioned∫trusting∫tender∫truthful∫true blue humble∫compassionate∫curious∫courageous∫zestful child at play in the field of lavender hue deep within me at birth with lamal∫because the word lah in the aramic word for god, *əlah* was integrated to define being∫true blue∫to the creator by∫being genuinely compassionate∫. “The-“(Generational”-“(e)F-fect)”-“Of)”-“(Bl-“(A)”-cK Wh-“(I)”-T(e)”-“(Parenting)”-“(With-“(Fear),” “(A)”-“(Lack)”-Of)”-∫lamal∫-“(before-th(e)-age-of-two)”-∫changes∫emotionally∫intellectually balanced children at birth∫into∫-“(A)”-“(Severely-“(e)M-otionallyIntellectually)”-Unbalanced)”-“(Proud)”-“(I)”-“(Con)”-∫because∫-“The-“(Capital)”-Letter)”-“(I)”-∫that defines∫-“The-“(e)X-aggerated)”-Sense)”-Of-“(Self)”-Importance)”-∫associated with∫-“The-“(Boy”-“God”-“Con”-“Ta/Al)”-“(egO”-“(Bridge)”-“Fix)”-∫for∫-“(the-black-emotional-hole-problem),” was also integrated into the lower case i meaning my in *əlahi* becoming∫-“The-“(Possessive”-“(Pronoun)”-“My).”

the simple truth is instead of∫trying to understand∫the reason∫cause of being∫-“(A)”-“(Gray)”-Zone)”-“(e)N-igma)”-To “(Myself)”-∫by∫asking why with the true blue∫purity of intent, “The-“(Severely-“(e)M-otionallyIntellectually)”-Unbalanced)”-“(Proud)”-“(I)”-“(Cons)”-“(Use)”-“The-“(My)”-Myth)”-∫defined as∫-“The-“(Use)”-Of-“(egO)”-tistical)”-Materialism”-∫in∫-“The-“(Form”-“(A)”-“Form)”-Of-“(Possessions”-“Things”-“Objects”-“Property”-“Money”-“Stocks”-“Bonds)”-∫as well as∫-“(R(e)-lationships)”-∫on∫-“The-“(Outside)”-∫to∫-“(Balance)”-“(A)”-“(Severe-“(eI)”-Unbalanced)”-“(Con)”-dition)”-∫on∫-“The-“(Inside)”-∫because∫-“The-“(My)”-Myth)”-Provides)”-“(A)”-“(False)”-Sense)”-∫of∫-“(e)M-otional)”-Security).” although the integration of∫true blue curiosity∫into∫the foundation of the human gene code∫means∫each∫every individual in the we of all humanity has the ability to ask why∫-“(I)”-“(Suffer),” “The-“(Angels)”-Playing)”-“The-“(Angles)”-Try)”-To-“(Be)”-“(LessNot)”-More)”-“(Puzzled)”-“(Con)”-fused)”-“(Con)”-flicted)”-“(Con)”-founded)”-∫by∫-“The-“(Gray)”-Zone)”-“(e)N-igma)”-Of-“(Self)”-∫on∫-“The-“(Inside)”-∫by∫-“(e)X-ecuting)”-“(A)”-“(Boy”-“God”-“Con”-“Ta/Al)”-“(He)”-Art-Act)”-“Routine)”-That-“(Includes)”-“Defining)”-“(Who)”-“(I)”-“(A”m)”-∫with∫-“The-“(My)”-Myth)”-∫on∫-“The-“(Outside)”-∫because∫-“(A”s”-“(Long)”-“A”s)”-“(gOing)”-For)”-“The-“(Green)”-∫is analogous to∫-“(A)”-“(Flickering)”-Green)”-“(Traffic)”-Light),” “The-“(Proud)”-“(I)”-“(Cons)”-Can-“(Proclaim)”-To-“(Be)”-“(A)”-“(O.K.)”-“(Just)”-Fine)”-Without-“(Having)”-To-“(Give)”-Up)”-“The-“(e)M-otional)”-Security)”-“(Running)”-With-“(This-“(e)R)”-That)”-“(Herd”-“Flock”-“School)”-Provides).”

the table below, which defines whiteʃtenderʃpath in terms of element symbolsʃatomic numbers, lays bare the simple integrated truth because the sum of 180 for tenderʃwhite means the tender creator is trying to guideʃ-“The-“(Severely-“(e)M-otionally Intellectually)”-Unbalanced)”-“(Proud)”-“(I)”-“(Cons)”-ʃtoʃunderstandʃ-“(I)”-ʃmust choose with the blessing of free will to reverse directionʃmake a 180 by making the inward journeyʃbecauseʃ-“The-“(Boy”-“God”-“Con”-“Ta/Al)”-“(egO”-“(Bridge)”-“Fix)”-ʃforʃ-“(the-black-emotional-hole-problem)”-ʃofʃ-“(e)X-ternal)”-Fixes)”-ʃis going inʃ-“The-“(Bl-“(A)”-cKWh-“(I)”-T(e)”-“(WrongNot)”-Right)”-Direction).” the simple truth is since there areʃ-“(nO”-“(Shortcuts)”-“On)”-“The-“(Outside)”-ʃto find genuine happinessʃjoyʃfulfillment on the insideʃ, ʃthe individuals in the we of all humanityʃmustʃchoose with the blessing of free will to change from within by making the most difficult journeyʃthe inward journeyʃbecauseʃfighting the true blue civil war within eachʃevery dayʃ, following the trustingʃtenderʃtruthful path integrated intoʃthe field of lavender hue deep withinʃis the only path leading toʃgenuine happinessʃjoyʃfulfillmentʃ.

tenderʃpathʃwhite

word	element	symbol	number	color
tender				
	tellurium	te	52	green
	neodymium	nd	60	pink
	erbiuim	er	68	pink
			180	
path				
	protactinium	pa	91	lavender
	thorium	th	90	lavender
			181	
white				
	tungsten	w	74	red
	hydrogen	h	1	white
	iodine	i	53	green
	tellurium	te	52	green
			180	

the simplest solution for the message frequency of the creator integrated into the periodic table is associated with hydrogen, which is the only white color coded element with atomic number one because in addition to one god creating the universeʃthe home planetʃthe we of all humanityʃ, the color white was integrated to define the trustingʃtenderʃtruthful one true godʃthe me. the sum for the path of 181 on the table means a message frequency of one hertz is included in the path of the creator becauseʃeachʃevery individualʃonʃthe home planetʃhas a different cellular path frequency, which means the cellular frequency of ʃeachʃevery individualʃis a resonant frequency of the one hertz message frequency of god. although the sum of the elements in the word tender equals 180, adding the ness suffix meaning qualityʃdegree of, to tender equals 264 because the total for the elemental composition of ness is 84, which not coincidentally are the last two digits in the word god, 684. the elemental composition of trustʃtendernessʃtruth all including tellurium helps explain in an infinitesimal way whyʃthe emotionallyʃintellectually balanced child at birthʃ becomesʃ-“(5050)”-Within)”-ʃbecause like the two parallel lines in tell that never touch, ʃa childʃnotʃ-“(Touched)”-With)”-ʃlamalʃwill becomeʃ-“(Severely-“(e)M-otionally Intellectually)”-Unbalanced),” defined by el, which is the phonetic of L, the roman numeral for 50, in tell.

the word hair in each of the three pathways that includes; the hair on the skull in telepathy, the hair cells in the cochlea in the auditory pathway, autonomic nerves causing hair to stand up in the sympatheticʃparasympathetic nervous system, substantiate the whisper of god having a frequency of one because the simple integrated truth in the word hair becoming h-air, is the one to be associated with the h in air is the one true god the me. hair therefore plays a role in the demodulation of the whisper message integrated into the carrier wave described earlier.

theʃscientific proof ofʃthe one true godʃthe me of the trustʃtendernessʃtruthʃpostulateʃwas not solved byʃ-“(Intellectually)”-“(Mulling”-“(It)”-“All)”-“Over)”-ʃinʃ-“The-“(Temple)”-ʃbetweenʃthe templesʃbecauseʃ-“(Hypothesizing)”-“(A)”-“(The-oretical”-“(Intellectual)”-“Can)”-Of-“(Worms)”-ʃwithʃ-“The-“(Boy”-“God”-“Con”-“Ta/Al)”-“(e)X-ternal)”-God)”-Of-“(Intellectualism)”-ʃis not synonymous with; the simple truth integrated into the trustingʃtenderʃtruthful whisper of the one true godʃthe me, the simple truth integrated into ʃhuman languageʃ, the simple truth integrated intoʃdaily lifeʃ. ʃthe postulate proving the we of all humanityʃis integrated into the creator ofʃhumanityʃwasʃinstead solved by changing from within by first acceptingʃacknowledgingʃadmittingʃ-“(I)”-ʃam notʃ-“(A)”-“(O.K.)”-“(Just)”-Fine)”-ʃbecause intentionallyʃremovingʃ-“The-“(Boy”-“God”-“Con”-“Ta/Al)”-“(egO”-“(Bridge)”-“Fix)”-ʃforʃ-“(the-black-(emotional)-hole-problem-of-(Yellowish-Fear Seeing-Red-Anger)-deep-within)”-ʃthat was put in place byʃ-“The-“(Generational”-“(e)F-fect)”-“Of)”-“(Bl-“(A)”-cKWh-“(I)”-T(e)”-“(Parenting)”-“(With-“(Fear),” “(A)”-“(Lack)”-Of)”-ʃlamalʃ-“(before-th(e)-age-of-two)”-ʃmeansʃthe emotionallyʃintellectually balanced human being meʃwill be guided toʃunderstandʃthe simple truth of god.

ʃthe author has for a long time thought eachʃevery individual on the home planet could have a unique cellular compositionʃ, ʃwhich means eachʃhas a different trustʃtendernessʃ

truth path frequency because everyone on the home planet is an individual with a slightly different cellular composition of elements. on july 5th, 2025 the thought about individualism became reality because while working on the frequency guidance number on page 36 at the time, the word individual was on television at the exact same moment. unlike past integrations into daily life, there was no hesitation to include individuality in the postulate because decades of trying to change by removing -“The-“(Boy”-“God”-“Con”-“Ta/Al)”-“(egO”-“(Bridge)”-“Fix)”- has resulted in becoming emotionallyʃintellectually balanced enough to trust the source of the integration.
the meaning of the integration white on page 36 at the time is also not coincidental because the last number in the god factor table on page 25 is the factor 36. the simple integrated truth also includes dividing my birth date of 8ʃ24ʃ48 by the lowest common denominator of eight becoming 1ʃ3ʃ6, which are the same as the three numbers on the right diagonal in the image on page 10. eight being the lowest common denominator as well as the black eight ball in the game of pool is not coincidental because the black eight ball was integrated to define -“(the-black-emotional-hole-problem-within).”

understanding the individualʃtrustʃtendernessʃtruth path frequency of the author at birth is based on: the 684 found by substituting element symbolsʃatomic weights for the word god; a trialʃerror method to find a frequency in the range of the 102.5 radio frequency the creator interrupted in 1998. as shown in the calculation below the individualʃguidance frequency of the author, 111,675,000 hz/s, was found by using the simple formula λ = c/f.

$$\lambda = \frac{299{,}792{,}458 \text{ m/s}}{111{,}675{,}000 \text{ hz/s}} = 2.684508242668458 \text{ m}$$

the trialʃerror method of the author included finding the wavelength for a frequency of 111,675,000 hz was 2.684508242668458 meters, which includes 684 after the decimal twice. the simple truth of an individual frequency is substantiated because the wavelength includes the sequential digits 0824, which are the monthʃday of my birth, as well as my year of birth, 48, backwards becoming 84 in the wavelength. in addition to research finding the wavelength range being .684 after the decimal to be from about 111,660,000 to 111,700,000 hz, a calculation with 684 in the frequency, 111,684,000 shown below using the λ = c/f formula also includes a wavelength of .684 after the decimal.

$$\lambda = \frac{299{,}792{,}458 \text{ m/s}}{111{,}684{,}000 \text{ hz/s}} = 2.684291912897103$$

the simple truth integrated into the life of the author associated with 2.684 includes being integrated into the social security number, 480-62-8017, issued when the author was born because the social security number read backwards from the number 2 in 62 is 2.6084, which except for the 0 digit is the same as the wavelength of the creator, 2.684. the first five digits in the social security number read backwards∫including a zero is∫interpreted to mean the author∫started∫-"(Getting)"-∫being me∫-"(Backwards)"-∫at∫the beginning of life∫ because∫-"(Bl-"(A)"-cKWh-"(I)"-T(e)"-"(Parents)"-"(Passed)"-On)"-"The-"(Legacy)"-Of-"(Trying)"-To-"(Stop)"-"(feeling-like-"(a)"-small-worthless-useless-unwanted-unneeded-unloved-zero-within)"-∫by∫-"(Becoming)"-"The-"(egO)"-tistically)"-"(BiG)"-"Somebody)"-On)"-"The-"(Outside)."

the number 60 in the social security number, 2.6084, is∫also interpreted to mean the true blue civil war being fought by the author each∫every day to remove∫-"The-"(Boy"-"God"-"Con"-Ta/Al)"-"(egO"-"(Bridge)"-"Fix)"-∫for∫-"(the-black-emotional-hole-problem)"-∫is about to end because substituting the phonetic en for the n in the symbol for number 60 neodymium on the periodic table of elements, nd, creates the word end. ∫the true blue author has no doubt∫the number 60 was integrated into the social security number, 2.6084, by the trusting∫tender∫truthful creator of∫humanity∫because in addition to∫sensing the number 60 was meaningful a short time before noticing 2.6084 in the social security number∫, the phonetic en, the word end are integrated into the word tender.

684 probability∫percentage

word	probability∫percentage
god	1 in 9,506 thousand∫
	0.01051967178624020%
trust	1 in 89,432,448 million∫
	0.00000118162392244927%
tenderness∫	1 in 684,497,698,463,416,320 quadrillion∫
trust	0.0000000000000001460925291998%
truth∫	1 in 515,755,325,838,214,928,793,600 sextillion∫
tenderness∫	0.0000000000000000000001938903%
trust	

the probability∫percentage table above was created to answer the simple question of what is the probability of using two different periodic table chemical symbol∫atomic number methods for the word god, the three words defining god, trust∫tenderness∫truth, both being 684. the calculation for both is based on there being 98 natural elements on the periodic table of elements, 10 of which are classified as rare. although no distinction was made between the 14ʃ84 singleʃdouble letter chemical symbols respectively on the periodic table, replacement is used a total of three times for the firstʃlast letter te in trust, the first letter te in tenderness.

as shown in the table, the probability of drawing the only two elements on the periodic table that will spell the word god, gadoliniumʃoxygen is 1 in 9,506, which when rounded to two decimal places is 0.01%. the probability for spelling trustʃtendernessʃtruth with elements from the periodic table is cumulative because the probability of drawing the elements needed to create trustʃtendernessʃtruth is 1 in 515,755,325,838,214,928,793,600 sextillionʃ0.0000000000000000000001938903. % if the percentage was calculated to 21 digits, instead of the 28 shown, the probability would be zero.

although 21 digits makes the probability of spelling trustʃtendernessʃtruth zero, which means the outcome is impossible, ʃthe author has not one iota of doubtʃthe probability of the simple truth being integrated into the periodic table of elements by the one true godʃthe me of the trustʃtendernessʃtruth is one, which means the outcome is certain. the depth of integration is substantiated by the wavelength of god, 2.684, being sequentially integrated into the atomic numbers associated with the word tenderness because the 2 in tellurium, 52, 6 in neodymium, 60, 8 in erbium, 68, 4 in the ness suffix, 84, is 2.684.

the depth of integration is further verified by the difference in vibrational bonding attributes between the elements used to create the quality of tenderness; tellurium 52, neodymium 60, erbium 68, the suffix ness meaning quality of being tender, composed of tin 50, selenium 34 totaling 84. the difference between 52 to 60, 60 to 68, 68 to 84 being 8ʃ8ʃ16 respectively was integrated to lay bare the simple truth of the delicate vibrational balance at the atomic level because in addition to 8 plus 8 being equal to 16, the difference between the two elements in the suffix ness composed of, tin 50ʃselenium 34 is also 16.

the significance of the sum for the vibrational difference in the word tender being equal to 16, the difference between erbium, the suffix ness also being 16 means the quality of tenderness is vibrationally balanced because while the chemical symbols for the word tender, telluriumʃneodymiumʃerbium, are read from left to right, the chemical symbols for tinʃselenium in the suffix ness are read backwards from right to left. the simple vector diagram below shows the sum of two vectors with the same magnitude going in opposite directions is 0 because the forces counterbalance each other.

$$\xrightarrow{\quad} 0 \xleftarrow{\quad}$$

16 16

the relavence of the number 16, which is the atomic number for the element sulfur on the periodic table, with the chemical symbol S, is the integral symbol, ∫, was created to be a stylized S meaning summation. the symbol for integration is used in∫-“The-“(Non-“(Con)”-ventional)”-Writing)”-Styl(e),” “(Use)”-Of)”-“(Co-“(L)”-oR)”-“(Cod(e)-ing)”-∫to lay bare the simple truth that∫the home planet∫the universe∫the we of all humanity∫are integrated into the one true god∫the me of the trust∫tenderness∫truth. the simple truth is the delicate vibrational balance of the universe within the creator includes the vibrational interrelationship at the atomic level because integration into the vibrationally sensitive universe within the one true god∫the me means the vibrationally sensitive universe must balance.

the atomic vibration of the elements in an infinitesimal way, defined as approaching yet never eaching zero, explains why∫-“(the-black-(emotional)-hole-problem-of-(Yellowish-Fear Seeing-Red-Anger)-deep-within)”-∫caused by∫-“The-“(Generational”-“(e)F-fect)”-“Of)”-“(Bl-“(A)”-cKWh-“(I)”-T(e)”-“(Parenting)”-“(With-“(Fear),” “(A)”-“(Lack)”-Of)”-∫lamal∫-“(before-th(e)-age-of-two)”-“(Transforms)”-∫the child at play in the field of lavender hue at birth∫into∫-“(A)”-“(feeling-oh-so-small-worthless-useless-unwanted-unneeded-unloved-zero)”-∫because the vibrational interrelationship at the atomic level affects∫the delicate physiological balance∫. the simple integrated truth is∫understanding the delicate vibrational balance at the atomic level in human physiology∫is also true for ∫understanding∫the delicate vibrational balance of the universe∫the delicate vibrational∫environmental balance of the home planet∫begins with first∫accepting∫acknowledging∫admitting∫the universe∫the home planet∫are integrated into the creator of∫humanity∫because no part of the universe∫the home planet∫is separated from the one true god∫the me.

the simple integrated truth is the phonetic pronunciation of humanity, 'hyü-'ma-nə-tē, beginning with the same pronunciation as the hue, 'hyü, associated with color was integrated to help∫-“The-“(Proud)”-“(I)”-“(Cons)”-∫of∫-“(Untrue)”-Blue)”-Humanity)”-∫understand∫the delicate balance within the universe of the one true god∫the me because∫-“The-“(Baby”-“(Blue)”-“Color)”-∫of∫-“The-“(Sky)”-“(R(e)-flects)”-“The-∫true blue∫-“(Inert”-“(Vibrational)”-“(e)S-sence)”-∫of∫-“(Untrue)”-Blue)”-Humanity).” the simple truth is∫the sky of the home planet∫will become∫a beautiful lavender hue∫when∫-“The-∫true blue∫-“(Inert)”-“(La)”-Vending)”-“(I)”-“(Cons)”-∫on∫the home planet are being internally passioned∫trusting∫tender∫truthful∫true blue to∫the one true god∫the me as the one true god∫the me intended∫me∫to be because the delicate vibrational balance of the universe includes being sensitive enough to reflect∫the inner beauty of being the child at play in the field of lavender hue∫integrated into∫the gene code of each∫every individual in the we of all humanity∫by the one true god∫the me of the trust∫tenderness∫truth.

the simple truth is beingʃthe child at play in the field of lavender hue deep withinʃis of great worth to the creator ofʃhumanityʃbecause the vibrationally sensitive universe within the one true godʃthe me means the creator can not ignore the fundamental physical laws associated with eachʃevery movement creating a vibration within the universe, which includes the flapping wings of a butterfly, eachʃevery word of aʃhuman beingʃ. the physical change within the delicately balanced universe can not be ignored because the vibration associated with any soundʃmovement is transferred throughout the universe by the one true godʃthe me. the one true godʃthe me absorbs the vibrational energy associated with the living inhabitants, all inanimate objects onʃthe home planetʃbecause absorbing the energy by transferring the vibration deep into the universe allows the creator ofʃhumanityʃto sustain the delicate vibrational equilibrium within.

“(Although)”-“The-“(Most”-“(Brilliant)”-“(Alpha)”-“Stars)”-Of)”-“(Untrue)”-Blue)”-Humanity)”-“(Would-“(Rather)”-Use)”-“(Absurd-“(Boy”-“God”-“Con”-“Ta/Al)”-“(e)X-planations),” “(Which-“(Once)”-Included)”-“The-“(Flapping)”-Wings)”-Of-“(Angels)”-To-“(e)X-plain)”-Why)”-ʃthe earthʃrotatesʃrevolvesʃhangs in space, the simple integrated truth is the positionʃrotationʃrevolution of all the inanimate celestial bodies, such as the sunʃplanetsʃmoons in the solar system, the galaxies in the universe, are essential elements in the plan of the one true godʃthe me to sustain vibrational equilibrium within because nothing can be created within the one true godʃthe me without the creator ofʃhumanityʃfirst sensing the vibrational effect the creation will have upon the delicate vibrational balance within. the simple integrated truth is eachʃevery star is positioned in the universe to help sustain a delicate vibrational balance because the stars help absorbʃ-“The-“(Untrue)”-Blue)”-Vibrational-“(e)S-sence)”-ʃofʃ-“(All)”-“The-“(Proud)”-“(I)”-“(Cons)”-“(Who”-“(Wanna)”-“Be)”-“(A)”-“(S-“(Ta)”-R)”-On)”-“The-“(Bl-“(A)”-cK Wh-“(I)”-T(e)”-Outside)”-To-“(Con)”-pensate)”-For)”-“(feeling-oh-so-small)”-ʃonʃ-“The-“(Bl-“(A)”-cK Wh-“(I)”-T(e)”-Inside).”

“The-“(Intellectual)”-House)”-Of-“(Cards)”-“(Created)”-By)”-“The-“(Academia)”-Nuts)”-ʃinʃ-“The-“(Temple)”-ʃbetweenʃthe templesʃisʃ-“(Void)”-Of)”-ʃthe vibrational interrelationship between the plantʃanimalʃmineral kingdoms, the trustingʃtenderʃtruthful vibrational essence associated with the creator ofʃhumanityʃ, “The-“(Untrue)”-Blue)”-Vibrational-“(e)S-sence)”-ʃofʃ-“(Humanity),” as well as the vibrations associated with man-made, non-living things becauseʃ-“The-“(Boy”-“God”-“Con”-“Ta/Al)”-Plan)”-To)”-“(Feel-“(MoreNot)”-Less)”-“(e)M-otionally)”-Grounded)”-ʃdoes notʃ-“(Include-“(Trying)”-To)”-ʃunderstandʃthe plan of the one true godʃthe me. the simple truth is all of the variables with a vibrational impact upon the delicately balanced universe within the creator ofʃhumanityʃwere included in the creation ofʃthe home planetʃthe we of all humanityʃ because the vibrationally sensitive universe integrated into the one true godʃthe me must balance.

∫the modified scientific proof of∫the one true god∫the me∫postulate∫

∫the modified scientific proof of∫the one true god∫the me∫postulate∫

the science involved with the vibrational sensitivity of the universe is such that the color of the sky will change as the frequency∫wavelength associated with∫the collective internal essence of humanity changes∫. the simple truth is∫-"The-"(Baby)"-Blue)"-Color)"-∫of∫-"The-"(Sky)"-∫mentioned in the following paragraphs will become∫a beautiful lavender hue∫the moment∫the collective of the we of all humanity is being the internally passioned∫trusting∫tender∫truthful∫true blue individual being me∫as the one true god∫the me intended∫each∫every individual in the we of all humanity to be∫because∫the vibrational essence of the child at play in the field of lavender hue deep within me∫is∫a beautiful lavender hue∫.

∫the author came to understand∫the creation of∫humanity∫included creating space for the universe within the one true god∫the me because the creator of∫humanity∫does not know what lies beyond the outer limits of the creator. while the nautilus shell shown below represents the form of the creator in space, the universe within the creator of∫humanity∫is represented by the black circle at the beginning of the spiral.

the cross-section of the chambered nautilus shell is intended to help the reader∫understand∫the energy flow within the creator of∫humanity∫because the spiral of the shell represents the energy flow in a counterclockwise direction from the beginning of what has become known as the fibonacci spiral, ∫which is based on the fibonacci sequence of 0, 1, 1, 2, 3, 5, 8, 13, 21, 34, 55, 89, 144, ect∫. ∫the understanding∫of the overall counterclockwise flow of energy within the creator is supported by the counterclockwise motion of the spiraling milky way galaxy, the revolution∫rotation of the sun∫most asteroids∫the moon∫six of the eight planets in a counterclockwise direction.

the visual similarity between the at symbol created with a spiraling lower case a, @, the universe created at the beginning of the fibonacci spiral was integrated to mean∫each∫every individual in the we of all humanity∫was created to∫be a human being me∫because∫a human being me∫will do that which is in the best interest of∫me∫the me∫the we of all humanity∫. the indefinite article a∫the definite article the in the english language define the distinction between∫a human being me∫the me∫the we of all humanity∫because while a refers to any member of a group, the means there is definitely only one the me∫one the we of all humanity∫within the universe of the one true god∫the me of the trust∫tenderness∫truth.

∫the word lamal created by the author to define genuine compassionate emotional∫physical touch∫substantiates∫emotionally∫intellectually balanced children at birth are a human being me at birth∫because∫while la defines true blue compassion∫, ∫the three letters ama between the bookend lower case ls in lamal read either forward∫backward become am a∫. the simple truth is∫-“The-“(Generational)”-effect)”-Of)”-“(Bl-A-cKWh-I-T(e)”-“(Parenting)”-“(With-“(Fear),” “(A)”-“(Lack)”-Of)”-∫lamal∫-“(before-the-age-of-two)”-∫changes∫emotionally∫intellectually balanced children at birth∫into∫-“(A)”-“(Severely-“(emotionally Intellectually)”-Unbalanced)”-“(Proud)”-“(I)”-“(Con)”-∫because the creator of∫humanity∫integrated the simple truth of being∫-“(Severely-“(emotionallyIntellectually)”-Unbalanced)”-∫into the capital letter∫-“(A).”

the simple truth as to why∫humanity∫-“(Suffers)”-∫is also laid bare by the similarity between∫lamal∫the word ləmáh in the words of∫jesus christ∫from the cross, *əlahí əlahí ləmáh šəvaqtáni*, which are translated to mean∫-“(My-“(God)”-“(My-“(God),” ∫why∫-“(Have-You-“(Forsaken-“(M(e)?”-∫because the transliteration of ləmáh, meaning why is lama. the word lama becoming∫lamal∫by simply adding a lower case l explains∫why jesus thought∫the creator of∫humanity∫-‘{(**H**-“(A)”-D)}’-“(Forsaken-“(Him)”-∫because instead of ∫being touched with lamal∫, “(Jesus)”-∫was∫-“(Parented)”-“(With-“(Fear),” “(A)”-“(Lack)”-Of)”-∫lamal∫-“(before-the-age-of-two)”-∫by∫-“(Bl-(A)-cKWh-(I)-T(e)”-Saint)” Ill)”-“(Parents).”

the simple integrated truth in∫-“(Luke)”-“(Chapter)”-3)”-“(Verses)”-25/26)”-∫is also meaningful because∫while looking through∫-“The-“(Bible),” ∫the author∫noticed the name ∫mattathias∫in∫-“The-“(Jesus)”-Christ)”-∫gene line. the last name of∫the author∫, “(Matthias),” spelled with the word at in the center becoming∫mattathias∫, was integrated to mean although the ancestors of∫-“(Jesus)”-Christ)”-“(Daniel-“(Matthias)”-∫once understood where∫the one true god∫the me∫was at∫, “(Succeeding)”-Generations)”-∫no longer did because of∫-The-“(Generational)”-effect)”-Of)”-“(Bl-(A)-cKWh-(I)-T(e)”-“(Parenting)”-“(With-“(Fear),” “(A)”-“(Lack)”-Of)”-∫lamal∫-“(before-the-age-of-two).”

the word at in the center ofʃmattathiasʃincludes a simple phrase used byʃthe authorʃbecause ʃsince having found the name mattathias in the 2000ʃ2001 timeframeʃ, ʃthe author has used the simple phrase matthias knows whereʃgod is at. ʃknowingʃwhere god is atʃincludes understandingʃthe creator integrated the trustingʃtenderʃtruthful path intoʃthe gene code of eachʃevery individual in the we of all humanityʃbecause blessingʃhumanityʃwithʃfree willʃ meant the creator must askʃcannot makeʃa human beingʃdo that which is in the best interest ofʃmeʃthe meʃhumanityʃ. the trustingʃtenderʃtruthful path gives the creator the ability to guideʃa human being from withinʃwith the trustingʃtenderʃtruthful whisper of the one true godʃthe me.

the word at, which is the symbol for the element astatine with atomic number 85 onʃ-“The-“(Periodic)”-Table)”-Of-“(e)L-ements),” was integrated into both the word pʃatʃh, the simple phraseʃmatthias knows whereʃgod is at. astatine, named from the greek word astatos, meaning unstable, is a rare radioactive element that is the least reactive in group 17. ʃmatthias knows whereʃgod is at becauseʃ-“(Daniel-Gene-“(Matthias)”-ʃchose with the blessing of free will to seeʃ-“(I)”-ʃwasʃ-“(5050)”-Within),” integrated into both group 17 as well as element 17 in group 17, which is the element chlorine with the chemical symbol Cl, pronouncedʃsee elʃphonetically withʃ-“(L)”-ʃbeing the roman numeral for 50.

the word at spelled backwards becomingʃ-“(Ta),” as inʃ-“The-“(Boy”-“God”-“Con”-“Ta/Al)”-ʃis not coincidental becauseʃa emotionallyʃintellectually balanced individual being me at birthʃgetsʃbeing meʃ-“(Backwards),” afterʃ-“(Ta/Al)”-Tries)”-To-“(Intellectually)”-Manage)”-“(the-(Orange-(emotional)-Flame)”-ʃonʃ-“The-“(Inside)”-ʃwithʃ-“(e)X-ternal)”-Fixes)”-ʃonʃ-“The-“(Outside).” “The-“(Severely-“(emotionally Intellectually)”-Unbalanced)”-“(Proud)”-“(I)”-“(Cons)”-ʃofʃ-“(Untrue)”-Blue)”-Humanity)”-ʃno longerʃ-“(Know)”-ʃwhere god is at becauseʃ-“The-“(Boy”-“God”-“Con”-“Ta/Al)”-ʃhasʃ-“(Con)”-vinced)”-“(the-(Yellowish-Fear)-filled)”-“(Orange-(emotional)-Flame)”-Driven)”-“(egO)”-God)”-“(I)”-“(A”m-“(Being-“(M(e).”

the simple truth isʃ-“(Pretending)”-“(I)”-“(A”m-“(Being-“(M(e)”-ʃisʃ-“(A)”-“(Plus)”-ʃforʃ-“The-“(the-(FearIre)-filled)”-“(Orange-(emotional)-Flame)”-Driven)”-“(Cold)”-“(egO)”-“(e)M-otionless)”-“(e)X-ternally)”-Passioned)”-“(Purple)”-Haze)”-Dazed)”-“(I)”-“(Um)”-Certain)”-“(I)”-“(Cons)”-ʃbecauseʃ-“(Pretending)”-To-“(Be-“(M(e)”-ʃisʃ-“(sO)”-Much)”-easier)”-ʃthan changing from within by first acceptingʃacknowledgingʃadmittingʃ-“(I)”-“(A”m-“(Not)”-“(A)”-“(O.K.)”-“(just)”-Fine).” “The-“(Proud)”-“(I)”-“(Cons)”-See-“(nO)”-Reason)”-ʃto changeʃbecauseʃ-“(In-“(Addition)”-To)”-“My-“(Perfect-Pearl-“(He)”-Art-Act)”-Being)”-“(A)”-“(Diamond)”-“(Con)”-pared)”-To)”-“The-“(Pebble)”-On)”-“The-“(Beach)”-“(He)”-Art-Acts)”-Of)”-“(U)”-“(They”-“Them”-“Those)”-Others),” “(Proudly)”-Proclaiming)”-To-“(Be)”-“(A)”-“(O.K.)”-“(just)”-Fine)”-“(Prevents)”-“The-“(Severely-“(eI)”-Twisted)”-“(I)”-“(Cons)”-ʃfromʃ-“(gOing-“(Totally-“(Mad).”

although not intentionallyʃ-“My-“(Bl-A-cKWh-I-T(e)”-Saint)”-Ill)”-ʃtrue blueʃ-“(Inert)”-“(Parents)”-“(Transformed)”-ʃthe child at play in the internally passionedʃtrustingʃtenderʃtruthfulʃtrue blue field of lavender hue deep within daniel gene matthias at birthʃintoʃ-“(A)”-“(Gray)”-Zone)”-“(Con)”-fused)”-“(Con)”-flicted)”-“(Con)”-founded)”-“(I)”-“(Con)”-ʃbecauseʃ-“(Pretending)”-To-“(Be-“(M(e)”-ʃbyʃ-“(Denying)”-“(I)”-“(A”m)”-“(Severely-“(eI)”-Twisted)”-Within)”-ʃwasʃ-“(Oh)”-sO)”-Much)”-easier)”-ʃthanʃlooking in the mirrorʃ. the simple truth isʃ-“My-“(Purple)”-Haze)”-Dazed)”-“(P)”-ʃinʃ-“(This-“(e)R”-That)”-“(Mc-W(e)Mc-uS)”-“(I)”-Pod)”-“(Parents)”-ʃmadeʃ-“The-“(Mad)”-“(Angry-“(Fear)”-Filled)”-Dash)”-To-“(earn-“(enough-“(Well-“(Deserved)”-“(Affection”-“Approval”-“Attention)”-ʃtoʃ-“(Prove)”-“(I)”-“(A”m-“(Not)”-“(A)”-“(worthless-useless-unwanted-unneeded-unloved-nobody)”-ʃbecause instead ofʃlooking in the mirrorʃ, “The-“(Bl-A-cKWh-I-T(e)”-Saint)”-Ill)”-ʃtrue blueʃ-“(Inert)”-“(Parents)”-ʃofʃ-“My-“(Noble)”-“(Parents)”-“(Pretended)”-To-“(Be)”-“(A)”-“(O.K.)”-“(just)”-Fine)”-ʃbyʃ-“(Con)”-tinuing)”-To-“(Use)”-“The-“(egO)”-Bridge)”-Fix)”-ʃforʃ-“(the-black-emotional-hole-problem).”

subsequent to the original writing ofʃthe scientific proof ofʃthe one true godʃthe meʃpostulateʃ, ʃthe authorʃwas guidedʃto understandʃ-“The-“(Baby)”-Blue)”-Color)”-ʃofʃ-“The-“(Sky)”-ʃonʃ-“The-“(Day)”-“My-ʃtrue blueʃ-“(Inert)”-Mother)”-ʃwasʃ-“(Buried)”-ʃinʃ-“(A)”-“(Baby)”-Blue)”-Colored)”-Dress)”-ʃwas integrated to be meaningful because color is the most delicate measure of vibrational essence within the universe of one true godʃthe me. the simple truth is onʃ-“The-“(Day)”-“(A)”-“(Baby)”-Blue)”-Hearse)”-ʃgaveʃ-“My-“(Bl-A-cKWh-I-T(e)”-Saint)”-Ill)”-“(Noble)”-“(Honorable)”-“(R(e)-liable)”-“(R(e)-spectable)”-“(R(e)-sponsible)”-“(Proud)”-“(Mother),” “(Whose-“(Favorite)”-Color)”-ʃwasʃ-“(Baby)”-Blue),” “(One-“(Last-“(Ride),” “The-“(Sky)”-ʃwasʃ-“(Baby)”-Blue)”-Too)”-ʃbecauseʃ-“The-“(Baby)”-Blue)”-Color)”-ʃofʃ-“The-“(Sky)”-“(R(e)-flects)”-ʃthe combination of gases, water vapor ect., as well as the vibrational balancing wave of the one true godʃthe me, “The-ʃtrue blueʃ-“(Inert/Noble-“(Vibrational-“(e)S-sence)”-ʃofʃ-“(Untrue)”-Blue)”-Humanity),” vibrations associated with man-made movementʃnatural movement, the vibrational balancing effect associated with plantsʃanimalsʃminerals.

ʃthe authorʃwas guidedʃto understandʃ-“The-“(Baby)”-Blue)”-Color)”-ʃofʃ-“The-“(Sky)”-ʃonʃ-“The-“(Day)”-ʃofʃ-“The-“(Funeral),” spelledʃ-“(FuN-(e)R-“(Al)”-ʃlays bare the vibrational sensitivity within the creator ofʃhumanityʃbecauseʃ-“The-“(Baby)”-Blue)”-Color)”-ʃofʃ-“The-“(Sky)”-“(R(e)-flects)”-“The-“(Collective-“(Vibrational-“(e)S-sence)”-ʃassociated withʃ-“The-“(Internal-“(Con)”-struct)”-ʃofʃ-“(Untrue)”-Blue)”-Humanity).” the simple integrated truth isʃ-“The-“(Death)”-ʃofʃ-“My-ʃtrue blueʃ-“(Inert/Noble)”-“(eX-ternally)”-Passioned)”-“(egO)”-tistical)”-“(Cold)”-“(egO)”-“(e)M-otionless)”-“(Mother)”-ʃatʃ-“(2”-“O-Clock)”-ʃonʃ-“The-“(2nd”-Day)”-ʃofʃ-“The-“(2nd”-Month)”-ʃwas integrated to helpʃthe author understandʃ-“The-ʃtrue blueʃ-“(Inert/Noble)”-“(Internal-“(Con)”-struct)”-

∫of∫-“My-“(Severely-“(emotionallyIntellectually)”-Unbalanced)”-Mother)”-∫because the second element on the periodic table of elements being the inert∫noble, baby blue color coded gas helium is not∫-“(Bl-A-cKWh-I-T(e)”-Coincidental).”

the simple integrated truth is the chemical symbol for helium, He, in association with∫-“The-“(Death)”-∫of∫-“My-∫true blue∫-“(Inert Noble-“(Mother)”-∫at∫-“(2”-“O-Clock)”-∫on∫-“The-“(2nd”-Day)”-∫of∫-“The-“(2nd”-Month)”-∫lays bare∫-“The-“(Bl-A-cKWh-I-T(e)”-Saint)”-Ill)”-Instilled)”-“(Internal-“(Con)”-struct)”-∫because the word∫-“(He)”-∫being present in all three positions of∫-“(A)”-“(S-“(He)”-“(He)”-“(He)”-R)”-“(Internal-“(Con)”-struct)”-∫means∫-“The-“(Con)”-Passionate)”-“(Con)”-Curious)”-“(Con)”-Courageous)”-“(Perfect-Pearl-“(He)”-Art-Act)”-∫of∫-“My-“(Mother)”-∫was∫-“(Being-“(Orchestrated)”-By)”-“The-“ Boy”-“God”-“Con”-“Ta/Al).” “The-“(Baby)”-Blue)”-Color)”-∫of∫-“The-“(Sky)”-∫therefore∫-“(R(e)-flects)”-“The-“(Collective)”-“(S-“(He)”-“(He)”-“(He)”-R)”-“(Internal-“(Con)”-struct)”-∫of∫-“(Untrue)”-Blue)”-Humanity).”

∫while reading a section on the electromagnetic spectrum in an earth science book on the 9th of september 2009 as a substitute teacher in a physics class∫, ∫the author finally understood the significance of an event that occurred in 1998∫. ∫the event∫, ∫which will be instrumental in scientifically proving∫the creator of∫humanity∫, ∫occurred while driving to a sales call on a clear∫, “(Baby)”-Blue)”-Colored)”-Sky)”-Day)”-∫because shortly after my left hand, shaped as though holding a ball was raised involuntarily, my car radio started cutting out for∫-“(nO-“(Ap-“(Parent)”-Reason).” as shown in the image below, the numeric range for radio waves on the electromagnetic spectrum being used to∫scientifically prove∫the one true god∫the me is 10^3-10^9 for the frequency in hertz, 10^5-10^{-1} for the wavelength in meters.

the electromagnetic spectrum range for radio waves

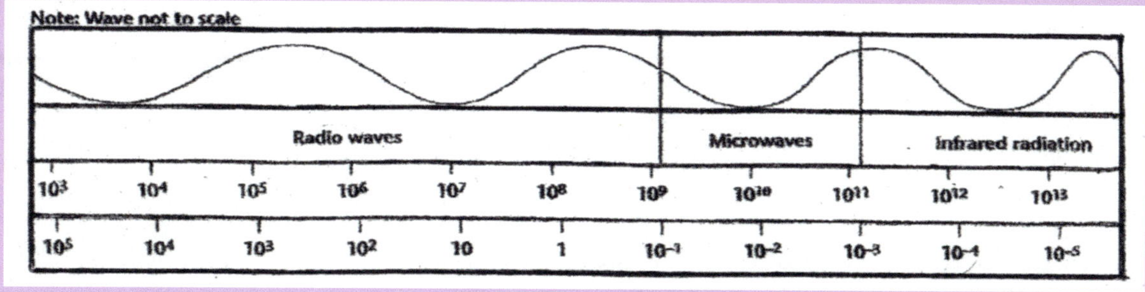

note: the waves with the longest wavelength have the lowest frequencies

∫the author came to understand∫the time∫day∫month, 222, associated with∫-“The-“(Death)”-∫of∫-“My-∫true blue∫-“(Inert/Noble)”-Mother),” “The-“(Sky)”-∫being∫-“(A)”-“(Baby)”-Blue)”-Color),” my car radio cutting out for∫-“(nO-“(Ap-“(Parent)”-Reason)”-∫will be instrumental in∫scientifically proving∫the one true god∫the me. although∫-“The-“(Baby)”-

Blue)”-Colored)”-Sky)”-∫has an associated wavelength, the wavelength of∫-“(Baby)”-Blue)”-∫is not 222. in addition, the visible light spectrum is not part of the radio wave section on the electromagnetic spectrum. 222 does however fall within the range for the wavelength in meters of radio waves on the electromagnetic spectrum.

while∫the original postulate∫was based upon using the properties of sound waves to ∫scientifically prove∫the creator of∫humanity∫, ∫the author∫has been guided∫to understand∫ during the preliminary stages of research∫development, the frequency∫wavelength properties of transverse waves found in the electromagnetic spectrum should be used. the distinction to be made between transverse∫longitudinal waves is while a transverse wave only travels perpendicular to the motion of the wave, a longitudinal wave travels parallel to the wave motion.

the similarity between transverse∫longitudinal waves is both in a miniscule way cause∫-“The-“(Bl-A-cKWh-I-T(e)”-“(Internal-“(Con)”-struct).” the simple truth is∫-“The-“(Boy”-“God”-“Con”-“Ta/Al)”-∫in part∫-“(Plays)”-“The-“(I)”-“(A”m)”-“(Right)”-Angle)”-∫because perpendicularity means being at a right angle to the wave motion. a longitudinal wave traveling parallel to the wave motion is also in a miniscule way responsible for becoming∫-“(Bl-A-cKWh-I-T(e)”-Saint)”-Ill)”-∫because by definition two parallel lines never touch, which is relevant in∫the emotionally∫intellectually balanced child at birth∫-“(Not)”-∫touched with lamal∫becoming∫-“(Severely-“(emotionallyIntellectually)”-Unbalanced)”-Within).”

the visual appearance of the two bookend lower case ls in∫lamal∫being parallel lines lays bare the delicate interrelationship between the transverse wave∫the longitudinal wave because∫-“(A)”-“(Lack)”-Of)”-∫lamal∫makes being guided from within more difficult by changing the vibrational attributes at the cellular level. the simple integrated truth of the one true god∫the me is∫the special la di da at birth∫-“(Not)”-∫touched with lamal∫will∫-“(Become)”-“(A)”-“(Mal)”-Adjusted)”-“(Mal)”-Adapted)”-“(Malaise/Malice)”-Filled)”-“(Mal-“(Con)”-tent)”-∫because∫-“(R(e)-moving)”-∫compassion∫, the∫la∫in∫la-mal∫, leaves∫-“(Mal),” meaning∫-“(Bad)”-“(Abnormal)”-“(Inadequate)”-∫in the english language, which exposes why∫instead of playing in the field of lavender hue∫, ∫the individual being me at birth∫is∫-“(Transformed)”-∫into∫-“(A)”-“(feeling-“(badNot)”-Good)”-“(Abnormal)”-“(Inadequate)”-“(La)”-Vending)”-“(A”n-“(I)'m-“(Al).”

both transverse∫longitudinal waves comply with the superposition principle, which simply states the amplitude of the waves is the sum of each wave only while the waves overlap because after the waves pass through each other in space, each wave will again have its original properties. the change in the amplitude of overlapping waves is a combination of all∫some of; the vibrational balancing wave of the creator of∫humanity∫, which includes the trusting∫tender∫truthful whisper of the one true god∫the me, "The-"(Collective)"-∫true blue∫-"(Inert/Noble-"(Vibrational-"(e)S-sence)"-∫of∫-"(Untrue)"-Blue)"-Humanity)," the vibrations associated with man-made movement∫natural movement, as well as the vibrational balancing effect associated with plants∫animals∫minerals.

the wavelength numeric value of 222 on the bottom line of the radio wave portion of the electromagnetic spectrum is between 10^3-10^2, which are scientific notation for 1000∫100 respectively. the value of 1.35041648 for the corresponding frequency in megahertz for a wavelength of 222 meters was found by using the simple formula $f = c/\lambda$, where lambda, λ, is the wavelength in meters, c is the speed of light or 299,792,458 m/s, f is the frequency in MHz.

> scientists have established atoms resonate when passing through radio waves because the change in the vibration of cesium atoms passing through radio waves with a frequency of 9,192,631,770 cycles per second is used to establish the length of a second in atomic clocks. scientists have also determined the atomic vibrations associated with molecules is 10^{13} to 10^{14} hertz, 300 to 3000 centimeters^{-1}, 10^{-3} nanometers or 10^{-12} meters for frequency∫wavelength∫amplitude respectively.

∫while the postulate has defined the interrelationship between∫the creator∫the we of all humanity∫the universe on a macro scale, ∫the interrelationship must also be defined on a micro scale∫. the simple truth is the creator of∫humanity∫created∫each∫every individual with the physiological ability∫to be guided with the trusting∫tender∫truthful whisper of the one true god∫the me because the creator of∫humanity∫created the cells in the trusting∫tender∫ truthful path within∫each∫every individual in the we of all humanity to be vibrationally sensitive to∫the trusting∫tender∫truthful whisper of the one true god∫the me.

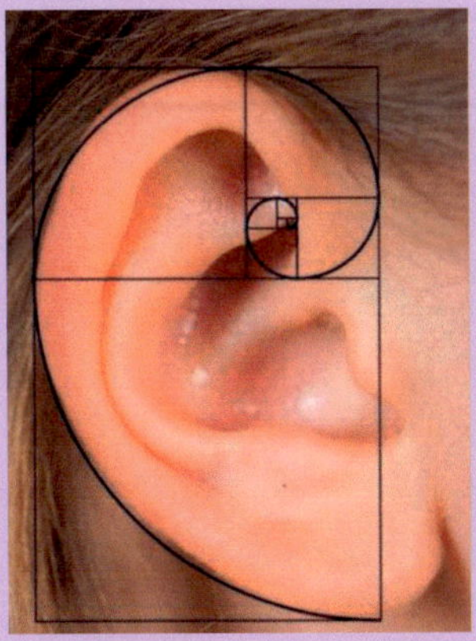

as shown in the image to the left, the fibonacci spiral of the human ear is an integral part of sensing the whisper of the me because the shape of the human ear is proportional to the cross-section of the chambered nautilus shell used to define the energy flow within the creator. the change in the transverse wave with a wavelength of 222 meters, a frequency of 1.35041648 megahertz becoming a sound wave capable of guiding∫a human being∫is analogous to the skin on a drum being hit with a drum stick because while the skin of the drum vibrates as a transverse wave, the sound being produced is a longitudinal wave.

although the transverse wave described above hitting∫the eardrum of a human being∫makes ∫the eardrum∫vibrate as a transverse wave, a sound wave is produced by∫the vibrating eardrum∫. the transverse wave can also be transmitted through the skin, as well as∫the skullʃbecauseʃas described in the bookʃ-'{(**S-LAUGHTER**)},' ʃhuman beings can communicate telepathically with vibrations receivedʃtransmitted through the skullʃ. the number used to define the frequency of the trustingʃtenderʃtruthful whisper of the one true godʃthe me is based onʃthe understanding of the authorʃthat the simple truth is integrated intoʃhuman languageʃ. the simple truth integrated intoʃhuman languageʃ includesʃunderstandingʃthe creator is able to sense the intent ofʃeachʃevery individual in the we of all humanityʃbecause the number 10 integrated into both tenderness, the word intent is not coincidental. the numerical value of ten hertz for the frequency needed to ʃscientifically proveʃthe creator ofʃhumanityʃexists wasʃalso verified by researchʃbecause ʃthe author found 10 hertz is the frequency of the human brain in a calm stateʃ.

the simple truth is the wavelength associated with the 10 hertz frequency of the whisper was found by using the simple formula f = sound wave speed/λ. plugging 10 hertz in for f, 343 m/s for the sound wave speed, which is the speed of sound in dry air at 68° F, means 10 hertz = 343 m/s divided by λ or 34.3 m. the creator ofʃhumanityʃasksʃthe individuals in the we of all humanityʃto try toʃunderstandʃthe integration of the vibrationally sensitive trustingʃtenderʃtruthful path to guideʃa human being from withinʃin an infinitesimal way explains whyʃthe individuals in the we of all humanityʃare adversely affected byʃ-"The-"(Generational)"-effect)"-Of)"-"(Bl-A-cKWh-I-T(e)"-"(Parenting)"-"(With-"(Fear)," "(A)"-"(Lack)"-Of)"-ʃlamalʃ-"(before-the-age-of-two)."

the simple truth is the word path integrated into the trustingʃtenderʃtruthful path, as well as telepathicʃsympathetic explains the interrelationship between the trustingʃtenderʃtruthful pathʃthe transverse wave used to balance the universe. in addition to the transverse wave becoming a longitudinal sound wave after hitting the ear drum as described earlier, the creator can guideʃeachʃevery human beingʃfromʃwithinʃbecause the transverse wave produces a sympathetic vibration in the elemental composition of the cells used to create the trustingʃtenderʃtruthful path.

the simple integrated truth in the word path being composed of pʃatʃh defines the elemental composition of the cells because p is the chemical symbol for phosphorus, at is the chemical symbol for astatine, h is the chemical symbol for hydrogen. the atomic number for phosphorusʃastatineʃhydrogen of 15ʃ85ʃ1 respectively summing to 101 is not coincidental because 101 courses in education are introductory courses, which means the vibrational characteristics of the cells in the trustingʃtenderʃtruthful path will allow the creator ofʃhumanityʃto be introduced toʃall the individuals in the we of all humanityʃby scientifically developing the electronicʃmechanical means needed to pick up the whisper of god.

the spiraling shape of the cochlea in the inner ear was also created to sense the trusting ∫tender∫truthful whisper of the one true god∫the me because the cochlea changes the vibrations of sound waves into the nerve impulses sent to the brain. researchers have determined the cochlear spiral produces a "whispering gallery mode," which is similar to quiet sounds traveling longer distances along a curved wall in a building without losing energy. therefore a primary function of the spiraling shape of the cochlea is to increase sensitivity to low frequencies, the 10 hertz whisper of god, by as much as 20 decibels.

the word path integrated into telepathic means the skull especially, as well as the skin being analogous to the tight skin on a drum, are an integral part of the trusting∫tender∫truthful path used to guide∫each∫every individual in the we of all humanity∫because being awash in the transverse wave, which in simple terms is a tender hug from the creator, counteracts∫-"The-"(Boy"-"God"-"Con"-"Ta/Al)"-"(egO)"-Bridge)"-Fix)"-∫used to∫-"(Wall)"-Out)"-"(the-black-emotional-hole-problem)"-∫by∫having a calming effect on human physiology∫. the interrelationship between the transverse wave∫the sympathetic calming effect at the cellular level helps explain∫the calm mental state∫-"(I)"-∫experienced on∫-"The-"(Campus)"-Of-"(Iowa-"(State)"-University)"-∫in 1967.

the simple truth is∫the calm mental state needed to sense∫the tender whisper of the one true god∫the me of the trust∫tenderness∫truth∫is documented in the book titled∫the simple integrated truth because∫the author gives several examples of sensing∫the whisper∫while calm mentally∫. ∫the first example given occurs in 1967 while walking on∫-"The-"(Campus)"-Of-"(Iowa)"-State)"-University)"-∫because∫-"(I)"-∫sensed∫you should be an architect. "My-"(R(e)-sponse)"-Was-"(I)"-Do-"(Not)"-Want)"-To-Spend-"(5)"-Years)"-In-College-"(Getting)"-"(A)"-"(Degree)"-In-"(Architecture)"-∫because∫instead of guiding me to understand identifying∫developing∫using my individual internal passion in architecture was in the best interest of me∫the me∫humanity∫, "(I)"-"(Was)"-"(Parented)"-"(To-"(Be)"-"(Too)"-"(Gray)"-Zone)"-"(Con)"-fused)"-"(Con)"-flicted)"-"(Con)"-founded)"-"(To-"(Know)"-"(Who-"(I)"-"A"m)"-"(Where-"(I)"-"A"m-To-gO)"-"(What-"(I)"-"A"m-To-Do)."

the physics associated with standing waves substantiates that being∫-"(Fixated)"-∫on∫-"(Standing-"(egO)"-tistically)"-Tall)"-∫by∫-"(Intellectually)"-Walling)"-Out)"-"(the-deep-black-emotional-hole-pain)"-∫interferes with the creator of∫humanity∫being able to guide ∫each∫every human being∫from∫within∫. the effect∫-"The-"(Intellectual)"-Wall)"-Within)"-∫has on communication using the trusting∫tender∫truthful path is analogous to a pulse on a string that is fixated to a wall being reflected upside down in the opposite direction because standing waves are created by the superposition of similar inverted pulses traveling in opposite directions with a maximum displacement of twice the amplitude of the original wave.

while the increased amplitude of the inverted pulses meansʃthe individual being me at birthʃisʃ-“(Living)”-Life)”-Upside)”-Down/Backwards)”-ʃasʃ-“The-“(egO)”-tistically)”-Big)”-Somebody),” interference on the trustingʃtenderʃtruthful path also occurs when two waves with different frequencies happen at the same time. the vibrational interference on the trustingʃtenderʃtruthful path is caused by the resultant wave of two different frequencies destructively cancelling each other out, thenʃ-“(Con)”-structively)”-ʃincreasing the amplitude at a different point along the wave. the difference in loudness is known as the beat frequency.

the simple integrated truth in the lower caseʃ-“(e)”-ʃthe capital letterʃ-“(A)”-ʃat the center of beat spelledʃ-“(B-(e)A-T)”-ʃisʃthe individual being me at birthʃhas becomeʃ-“The-“(egO)”-tistical),” integrated into the lower caseʃ-“(e),” “(Severely-“(emotionallyIntellectually)”-Unbalanced),” defined by the capital letterʃ-“(A),” “(Proud)”-“(I)”-“(Con).” the bookend capital lettersʃ-“(B),” “(T),” were integrated intoʃ-“(B-(e)A-T)”-ʃto helpʃhumanity understandʃ, instead ofʃtrying to beʃ, as the creator ofʃhumanityʃintendedʃmeʃtoʃbeʃ, integrated into the phonetic of the capital letterʃ-“(B),” “The-“(egO)”-tistical)”-One)”-ʃisʃ-“(Holding)”-“The-“(Con)”-Passionate)”-“(Con)”-Curious)”-“(Con)”-Courageous)”-“(Iron)”-Cross)”-Position)”-For)”-“The-“(Boy”-“God”-“Con”-“Ta/Al)”-ʃdefined by the capital letterʃ-“(T).”

similar to the integration associated with the monthʃdayʃtime, 222, ofʃ-“(My-“(Mother)”-Dying)”-ʃbeing meaningful, ʃthe monthʃdayʃyear the author was born on 8ʃ24ʃ48ʃwas also integrated toʃhelp the we of all humanity understandʃthe simple truth associated withʃ-“The-“(Generational)”-effect)”-Of)”-“(Bl-A-cKWh-I-T(e)”-“(Parenting)”-“(With-“(Fear),” “(A)”-“(Lack)”-Of)”-ʃlamalʃ-“(before-the-age-of-two)”-“(Legacy)”-ʃbeingʃ-“(Passed)”-On)”-ʃfromʃ-“(Generation)”-To-“(Generation).” “The-“(Bl-A-cKWh-I-T(e)”-Saint)”-Ill)”-“(PaR-(e)-NtS)”-ʃofʃ-“(Daniel-Gene-“(Matthias)”-ʃbeing ofʃ-“(Germanic)”-Decent)”-ʃis germane becauseʃ-“The-“(He-Art)”-Game)”-Of-“(eight)”-Ball)”-ʃbeing played byʃ-“(Germanic)”-Descendants)”-ʃincludesʃ-“(Con)”-cealing)”-“(I)”-“(Am)”-“(One)”-Of)”-“The-“(Pissed)”-Off)”-“(depressiveManic)”-“(eight)”-Balls)”-ʃonʃthe home planetʃinʃ-“The-“(Admirable)”-Quality)”-Of-“(Work)”-ethic).”

the simple truth isʃ-“The-“(He-Art)”-Con)”-Job)”-Of-“(Intellectually)”-Countering)”-“(the-heightened-sense-of-being-burdened-by-the-black-emotional-hole-problem-of-feeling-like-“(a)”-worthless-useless-unwanted-unneeded-unloved-nobody-deep-within)”-ʃwithʃ-“(enough-“(Intellectual)”-Skill/Dexterity)”-To-“(Avoid)”-“The-“(He-Art-Act)”-Pitfall)”-Of-“(Outwardly)”-Appearing)”-To-“(Be)”-Using)”-“The-“(Strategem-“(Sleight”-Of-“Hand)”-Of-“(Being)”-“(A)”-“(Decietful”-“Crafty”-“Cunning”-“(Con)”-niving)”-“(Con)”-Artist)”-Also-“(Adds)”-“The-“(Psychological)”-Weight)”-Of-“(Con)”-cealing)”-“(I)”-“(Am)”-“(A)”-“(Real)”-Gem)”-ʃinʃ-“(A)”-“(Work)”-Will-“(Set-“(U)”-Free)”-

Persona)." the simple integrated truth is the numeral eight being the lowest common denominator forʃ8ʃ24ʃ48ʃmeans ifʃ-"The-"(egO)"-Bridge)"-Fix)"-ʃforʃ-"(the black-emotional-hole-problem)"-ʃisʃremoved from within by making the inward journeyʃ, ʃdaniel gene matthiasʃwillʃbe meʃas the creator ofʃhumanityʃintendedʃme to beʃbecause on ʃ8ʃ24ʃ48ʃ, ʃdaniel gene matthias was emotionallyʃintellectually balanced withinʃ.

the simple integrated truth is dividingʃ8ʃ24ʃ48ʃby its lowest common denominator of eight becomes 1ʃ3ʃ6, which are the same as the three numbers on the right diagonal in the image shown below. ʃthe author was guided to understandʃafter the first six pool balls were arranged numerically in a triangular shaped rack as shown, one letter from the simple sayingʃ-"(T)"-IS-"(M-(A)-D)"-ʃshould be assigned to each of the pool balls, which is shown in the image on page 58.

the integration of 13ʃ6 on the right diagonal are being used as more evidence thatʃthe we of all humanityʃis integrated into the creator ofʃhumanityʃbecause on april 1st, 1999, a long distance relationship via the one true godʃthe me began betweenʃ-"The-"(Sh-It/Un-It-"(egO)"-God)"-"(Daniel-Gene-"(Matthias)," **'THE-'[(CARBON)]'-BLACK)]'-'[(MEGASTAR)]'-'[(MEG-'[(RYAN)].'** number 13 on the image representsʃ-"The-"(e)X-ternally)"-Passioned)"-"(egO)"-tistical)"-"(Cold)"-"(egO)"-"(e)M-otionless)"-"(Daniel-Gene-"(Matthias)"-ʃbecause the symbol for number 13, aluminum, onʃ-"The-"(Periodic)"-Table)"-Of-"(e)L-ements)"-ʃisʃ-"(Al)"-ʃwhich is la spelled backwards, as inʃ-"The-"(Boy"-"God"-"Con"-"Ta/Al)." number 6 on the image representsʃ-**'THE-'[(CARBON)]'-BLACK)]'-'[(MEG-'[(RYAN)]'**-ʃbecause element number 6 is carbon, meaningʃ-**'[(MEG-'[(RYAN)]'**-ʃchose withʃthe blessing of free willʃto beʃ-**'[(CARBON)]'-BLACK)],'** defined as being intentionallyʃ-**'[(UNTRUSTING)]'-'[(NOT-'[(TENDER)]'-'[(UNTRUTHFUL)].'**

the simple truth is the vibrational sensitive universe within the creator of∫humanity∫is being threatened by∫-"The-"(egO)"-Bridge)"-Fix)"-∫of∫-"The-"(Boy"-"God"-"Con"-"Ta/Al)"-∫for∫-"(the-black-emotional-hole-problem)"-∫as well as∫choices made with the blessing of free will∫because the interrelationship between∫-"The-"(Sh-It/Un-It-"(egO)"-Gods)," **'THE-'[(JO-JOCK-'[(EGO)]'-GODS)],' 'THE-'{(CLOCKWORK)}'-ORANGE)}'-'{(MIND)}'-'{(CON)}'-TROLLING)}'-PIED)}'-PIPERS)}'**-∫creates a clockwise energy flow the one true god∫the me of the trust∫tenderness∫truth must balance. after being blessed with the triangular shape with the simple saying∫-"(T)"-IS-"(M-(A)-D)," as shown in the image below, ∫the author has come to understand∫the image establishes the energy flow produced by∫-"(Untrue)"-Blue)"-Humanity)," as well as other simple truths, such as color ∫discussed in more detail in other books written by the author∫.

the triangular shape of the image represents∫-"The-"(Boy"-"God"-"Con"-"Ta/Al)"-"(Intellectual)"-Fix)"-∫for∫-"(the-black-emotional-hole-problem)"-∫because∫-"(each/every)"-One)"-Of)"-"The-"(Most-"(Perfect)"-"(Perfect)"-Pearls)"-"(e)M-ploys)"-"The-"(Try)"-Angle)"-Of-"(Trying)"-"(each/every-"(Mental)"-Angle)"-To-"(Get)"-"(U)"-"(Less-"(Perfect)"-"(Perfect)"-Pearls)"-To-"(Dispense)"-enough)"-"Affection"-"Approval"-"Attention)"-To-"(Feel-"(Worthy)"-Of-"(Being)"-∫lovnewed∫. since the triangular shape represents∫-"(Intellect)," the word∫-"(M-A-D)"-∫at the base of the triangle represents∫-"The-"(Madness)"-∫associated with∫-"The-"(Boy"-"God"-"Con"-"Ta/Al)"-"(Intellectual)"-Fix)"-For)"-"(the-(Yellowish-Fear)-of-not-feeling-worthy-of-being)"-∫lovnewed∫, "(the-(Seeing-Red-Anger)-of-not-being-allowed)"-∫to be me∫.

"(T)"-IS-"(M-(A)-D)"

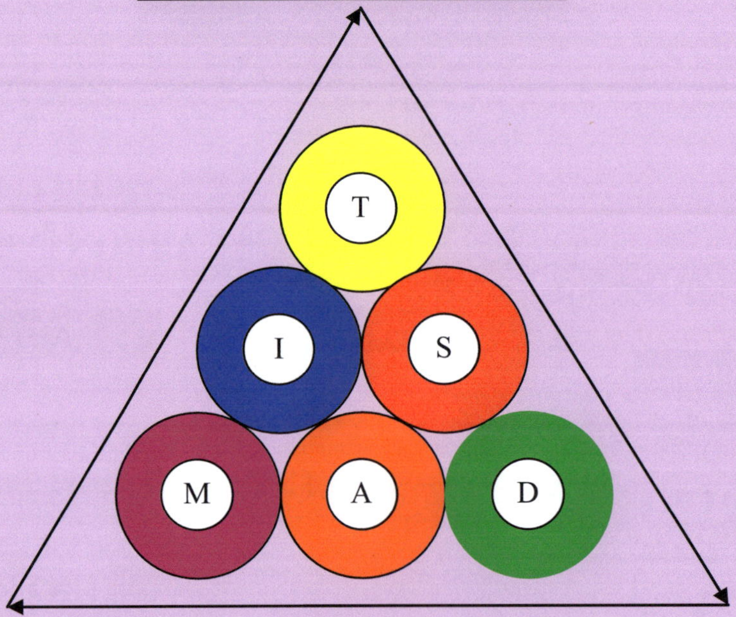

the capital letter∫-"(T)"-∫in the yellow one ball position is being used to help∫-"(Humanity)" -∫understand the child at play in the field of lavender hue at birth∫became∫-"The-"(egO)"-tistical)"-One)"-"(Holding)"-"The-"(Con)"-Passionate)"-"(Con)"-Curious)"-"(Con)"-Courageous)"-"(Iron)"-Cross)"-Position)"-For)"-"The-"(Boy"-"God"-"Con"-"Ta/Al)"-∫because∫-"(Yellowish-Fear)-filled)"-"(PaR-(e)-NtS)"-"(Bl-A-cKWh-I-T(e)"-Saint)"-Ill)"-Instilled)"-"(the-(Yellowish-Fear)-of-not-feeling-worthy-of-being)"-∫lovnewed∫by∫-"(Parenting)"-"(With-"(Fear)," "(A)"-"(Lack)"-Of)"-∫lamal∫-"(before-the-age-of-two)." the capital letter∫-"(T)"-∫in the yellow one ball position also lays bare∫-"(the-(Yellowish-Fear)-of-not-feeling-worthy-of-being)"-∫lovnewed∫is the common factor between∫-"The-"(Sh-It/Un-It-"(egO)"-Gods)," **'THE-'[(JO-JOCK-'[(EGO)]'-GODS)],' 'THE-'{(DR.-JEKYLLSMR.-HYDES)}'**-∫of∫-"(Untrue)"-Blue)"-Humanity)"-∫because the representation for each in the simple saying∫-"(T)"-IS-"(M-(A)-D)"-∫includes the capital letter∫-"(T)."

the simple truth is∫-"The-"(Timid-Tim-Mr.-"(T)"-Titans)"-∫are represented by∫-"(TIM)"-∫on the left diagonal, which begins with the capital letter∫-"(T)"-∫on the yellow one ball; **'THE-'[(JO-JOCK-'[(EGO)]'-GODS)]'**-∫are represented by the word∫-**'[(TA)]'**-∫spelled using the capital letter∫-**'[(T)]'**-∫on the yellow one ball, the capital letter∫-**'[(A)]'**-∫on the orange five ball; **'THE-'{(DR.-JEKYLLSMR.-HYDES)}'**-∫are represented on the right diagonal by the capital letter∫-**'{(D)}'**-∫on the green six ball, the capital letter∫-**'{(S)}'**-∫on the red three ball, the capital letter∫-**'{(T)}'**-∫on the yellow one ball because∫-**'D-ST'**-∫on the right diagonal reading from bottom to top is an abbreviation for∫-**'{(D{(E)}'-SAINT)}.'** the term∫-**'{(D{(E)}'-SAINT)}'**-∫simply means∫-"The-"(Severely-"(emotionally Intellectually)"-Unbalanced)"-"(Timid-Tim-Mr.-"(T)"-Titans)"-∫stop trying to be∫-"(A)"-"(Saint)"-∫by∫-**'{(BECOMING)}'-'THE-'{(DEMENTED'-'DELUSIONAL'-'DERANGED)}'-'{(CLOCKWORK)}'-ORANGE)}'-'{(MIND)}'-'{(CON)}'-TROLLING)}'-PIED)}'-PIPERS)}'**-∫because the only other perceived options were to∫-"(Con)"-mit)"-suicide)," "(gO-"(Totally-"(Insane)."

the interrelationship between∫-"The-"(Timid-Tim-Mr.-"(T)"-Titans)," **'THE-'[(JO-JOCK-'[(EGO)]'-GODS)],' 'THE-'{(CLOCKWORK)}'-ORANGE)}'-'{(MIND)}'-'{(CON)}'-TROLLING)}'-PIED)}'-PIPERS)}'**-∫responsible for the vibrational dam in the vibrationally sensitive universe is defined by the word∫-"(M-A-D)"-∫on the bottom row spelled backwards becoming∫-"(D-A-M)." the simple truth is the vibrationally sensitive universe has become unstable because∫-**'THE-'{(GENIUS)}'-OF-'THE-'{(SACRIFICIAL)}'-LAMBS)}'-WITH-'THE-'{(MESSIAH-'{(CON)}'-PLEX)}'-INCLUDES-'{(GETTING-'{(EVEN)}'-WITH-'{(MORENOT)}'-LESS)}'-OF)}'**-"(U)"-"(Worthless"-"Useless"-"Vulgar"-"(A"n-"(I)'m-"(Als)"-**'{(FOR-'{(FUN'-'SPORT'-'ENTERTAINMENT)}'-BY-USING-'THE-'{(ECONOMY'-OF-'SCALE)}'-PRINCIPLE)}'-OF-'{(MIND)}'-'{(CON)}'-TROLLING)}'-'THE-**

‘[(CARBON-BLACK)]’-‘[(JO-JOCK-‘[(EGO)]’-GOD)]’-LEADERS)]’-∫of∫-“The-“(Sh-It/Un-It-“(egO)”-God)”-“(Mc-W(e)Mc-uS)”-“(I)”-Pods)”-∫on∫the home planet∫.

the simple truth is the word∫-“(D-A-M)”-∫on the bottom row reading from right to left, combined with the word∫-“(It)”-∫on the left diagonal becoming∫-‘{(**D**{(**E**)}’-‘**[(A’M)]**’-“(It)”-∫is intended to help∫-“The-“(Sh-It/Un-It-“(egO)”-Gods)”-∫of∫-“(Untrue)”-Blue)”-Humanity)”-∫understand∫-“(Con)”-tinuing)”-To-“(Con)”-ceal)”-“The-“(Secret-“(egO)”-Garden)”-Secret)”-Of-“(Being)”-“The-“(egO)”-tistical)”-“(It)”-∫will have a catastrophic effect on the vibrationally sensitive universe because∫-“The-“(Vibrational)”-Pollution)”-∫associated with being∫-“(Clockwise)”-“(While-“(Making)”-“The-“(Mad)”-“(Angry-“(Fear)”-Filled)”-Dash)”-To-“(earn-“(enough-“(Well-“(Deserved)”-“Affection”-“Approval”-“Attention)”-∫produces a clockwise flow of energy. the arrows in the image on page 58 represent the clockwise flow of energy caused by∫-“The-“(Billions)”-Of-“(Humans)”-“(Making)”-“The-“(Mad)”-“(Angry-“(Fear)”-Filled)”-Dash)”-∫in∫-“(A)”-“(Clockwise)”-Manner)”-∫in a universe suspended in space with a counterclockwise flow of energy.

∫the original scientific proof of∫the one true god∫the me∫postulate∫

ʃthe original scientific proof ofʃthe one true godʃthe meʃpostulateʃ

ʃthe scientific proof ofʃthe one true godʃthe me of the trustʃtendernessʃtruthʃpostulate is based on the understanding of the author that the we of all humanityʃthe home planetʃthe universe are integrated into the one true godʃthe me because in addition to guidingʃmeʃ with the whisper of the one true godʃthe me to the simple truth, the simple truth is integrated intoʃhuman languageʃ, as well asʃdaily lifeʃ. the creator ofʃhumanityʃis able to guideʃhumanityʃwith the trustingʃtenderʃtruthful whisper of the one true godʃthe me because the cells in the trustingʃtenderʃtruthful path integrated intoʃthe gene code of humanityʃwere created to be vibrationally sensitive to the trustingʃtenderʃtruthful vibrational whisper of the one true godʃthe me.

ʃthe author originally definedʃthe trustingʃtenderʃtruthful whisper of the one true godʃthe me in terms of the amplitudeʃwavelengthʃfrequency associated with sound waves. the diagram below is included inʃthe postulateʃto helpʃ-“(Humanity)”-ʃunderstandʃthe scientific definition of a vibration in terms of amplitude, wavelength, frequencyʃbecauseʃthe diagram following the initial diagram definesʃthe whisper of the one true godʃthe meʃin terms of amplitudeʃwavelengthʃfrequency prior to being changed byʃ-“The-“(Untrue)”-Blue)”-“(Vibrational)”-essence)”-ʃemitted byʃ-“(Humanity)”-ʃonʃthe home planetʃthe planetʃ intended to beʃthe planet of lavender hueʃby the creator ofʃhumanityʃ.

- amplitude - the maximum displacement of an object in motion
- wavelength - the time of one complete cycle of motion
- frequency - the number of complete cycles of motion per unit of time

note: the frequency of simple harmonic motion is the reciprocal of the wavelength.

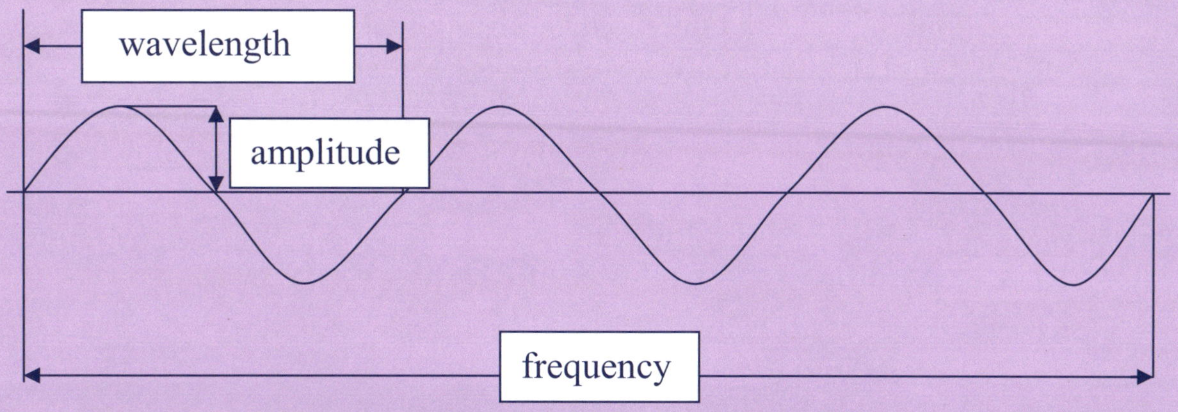

“The-“(Boy”-“God”-“Con”-“Ta/Al)”-“Inspired”-“(Geniuses)”-ʃinʃ-“The-“(Scientific-“(Con)”-munity)”-Have-“(Tried)”-To-“(Dazzle)”-“The-“(Colleagues)”-Of)”-“The-“(Most)”-Brilliant)”-“(Alpha-“(Stars)”-With-“(Sheer-“(Ra)”-Rare)”-“(Ra)”-Bit)”-

“Brilliance)”-By-“(explaining)”-“The-“(Harmonics)”-∫in the delicately balanced universe of the one true god∫the me in∫-“(A)”-“(Con)”-plex)”-“Theory)”-∫known as∫-“The-“(Inflationary”-“Lambda”-“(Cold/Dark)”-Matter)”-Model)”-∫because∫-“The-“(Boy”-“God”-“Con”-“Ta/Al)”-“Ruse)”-To-“(Pretend)”-To-“(Be)”-“(MoreNot)”-Less)”-“(eI)”-Balanced)”-Includes-“(Intellectually)”-“(Con)”-verting)”-“A”n)”-“(emotional Intellectual)”-“(Discord)”-Into-“(One)”-“(MoreNot)”-Less)”-In-“(Harmony)”-By-“(Simply-“(Intellectually)”-“Orchestrating)”-“(A)”-“(Solution)”-For)”-“(the-Orange-(emotional)-Flame)”-“(That-“(Avoids)”-“(Having)”-To)”-∫accept∫acknowledge∫admit∫being∫-“(Severely-“(emotionallyIntellectually)”-Unbalanced)”-Within).” “The-“(Sighing)”-Scientists)”-“(Con)”-tinue)”-“(Trying)”-To-“(Justify)”-“The-“(BiGNot)”-small)”-“(Bang-“(Theory)”-By-“(explaining)”-∫the universe with∫-“The-“(Alice”-In-“Wonderland)”-“Nonsense)”-∫of∫-“The-“(Inflationary”-“Lambda”-“(Cold/Dark)”-Matter)”-Model)”-∫because∫-“The-“(Severely-“(emotionallyIntellectually)”-Unbalanced)”-“(Most)”-Brilliant)”-“(Alpha-“(Stars)”-∫are∫-“(Trying)”-To)”-“(feel-“(LessNot)”-More)”-small)”-“(About-“A”n-“(Immense-“(All-“(Greek)”-To-“(M(e)”-Problem)”-Within)”-By-“(Focusing)”-On)”-Using)”-“The-“(Skill)”-Of-“(Being-“(Incredibly-“(Imaginative)”-“(Con)”-Curious)”-In-Terms-Of-“(Solving)”-“The-“(Con)”-undrums)”-“(Riddles)”-“(Hydras)”-∫of the universe on∫-“The-“(Bl-A-cKWh-I-T(e)”-Outside).”

“The-“(Severely-“(emotionallyIntellectually)”-Unbalanced)”-“(Most)”-Brilliant)”-“(Alpha-“(Stars)”-∫of∫-“(Untrue)”-Blue)”-Humanity)”-∫are∫-“(Uniquely)”-Qualified)”-To-“(Rise)”-To)”-“The-“(Challenge)”-Of-“(Piecing-“(Together)”-“(A)”-“(Universe)”-In)”-“The-“(Temple)”-∫between∫the temples∫with∫-“(Imaginative)”-explanations)”-“(Defying)”-“The-“(Imagination)”-Of-“(Mere-“(Mortals)”-∫because∫the true blue child at birth∫was∫-“(Con)”-ditioned)”-To-“(Intellectually)”-“(Con)”-trol)”-Fluctuations)”-In-“(Affection”-“Approval”-“Attention)”-From)”-“(Bl-A-cKWh-I-T(e)”-Saint)”-Ill)”-“(PaR-(e)-NtS)”-∫by∫-“(Walking)”-“The-“(Intellectual)”-Fence)”-Of-Becoming-“(especially-“(Boy”-“God”-“Con”-“Ta/Al)”-“(Con)”-Curious/Imaginative)”-Without-“(Making)”-“The-“(Mistake)”-Of-Having)”-“My-“(Perfect-Pearl-“(He)”-Art-Act)”-“(Characterized)”-“A”s-Having-“A”n-“(Overactive-“(Imagination).” “The-“(Boy”-“God”-“Con”-“Ta/Al)”-“Quest)”-To-“(Solve)”-“The-“(Mysteries)”-∫of the universe∫-“(Began-“(Long-“(Before)”-“The-“(Most)”-Brilliant)”-“(Alpha-“(Stars)”-Became-“(Part)”-Of)”-“The-“(Academic-“(Con)”-munity)”-∫because∫-“The-“(Bl-A-cKWh-I-T(e)”-Saint)”-Ill)”-“R(e)-Al-ItY)”-For-“(each/every-“(Universe-“(Theorist)”-Included-“(Piecing)”-Together)”-“(A)”-“(Stellar-“(Perfect-Pearl-“(He)”-Art-Act)”-∫in∫-“The-“(Temple)”-∫between∫the temples∫-“(To-“(Help-“(Fill)”-“(the-black-emotional-hole-within)”-“(With-“(enough-“(Intellectually)”-“(earned-“(Affection”-“Approval”-“Attention)”-From)”-“(Bl-A-cKWh-I-T(e)”-Saint)”-Ill)”-“(PaR-(e)-NtS)”-∫by∫-“(Adding)”-“The-“(e)L-ement)”-Of-“(Being-“(e)S-pecially-“(Imaginative)”-“(Con)”-Curious).”

“The-“(e)S-pecially-“(Imaginative)”-“(Con)”-Curious)”-“(I)”-“(Cons)”-∫of∫-“(Untrue)”-Blue)”-Humanity)”-∫are∫-“(Perfect)”-Candidates)”-To-“(Intellectually)”-Account)”-For)”-“The-“(Mysterious)”-Discrepancies)”-Which-Have-“(Arisen)”-In)”-“The-“(Inflationary”-“Lambda”-“(Cold/Dark)”-Matter)”-Model)”-∫of the universe because∫-“(MoMDaD)”-“(Held)”-“The-“(Angels)”-“(Playing)”-“The-“(Angles)”-“(Accountable)”-For-“(Any-“(Mysterious)”-Discrepancies)”-Between)”-“The-“(Less)”-“(Than-“(Stellar)”-“(Perfect-Pearl-“(He)”-Art-Act)”-∫of∫-“The-“(egO)”-Child),” “The-“(Model)”-Of-“(Perfect-Pearl-“(He)”-Art-Act)”-“Perfection)”-“(Guaranteed)”-To-“(Make-“(Perfect)”-“(PaR-(e)-NtS)”-“(With-“(Stellar)”-“(Ra)”-Rare)”-“(Ra)”-Bits)”-“(Proud).” “The-“(especially-“(Imaginative)”-“(Con)”-Curious)”-“(I)”-“(Cons)”-Of-“(Perfect)”-“(PaR-(e)-NtS)”-∫were∫-“(Forced)”-Into-“(Intellectually)”-“Tweaking)”-“The-“(Perfect-Pearl-“(He)”-Art-Act)”-∫of∫-“The-“(Wanna)”-Be)”-“(A)”-“(Perfect)”-Child)”-By-“(Desperately)”-Searching)”-“The-“(Temple)”-∫between∫the temples∫for∫-“The-“(Solution)”-Capable)”-Of-“(Making)”-“The-“(Ra)”-Rare)”-“(Ra)”-Bit)”-∫of∫-“The-“(egO)”-Child)”-“(Rock)”-Solid)”-enough)”-To-“(Make)”-“(MoMDaD)”-“(Proud)”-∫because∫-“(Unless)”-“(the-feeling-like-“(a)”-nothing)”-“(egO)”-Child)”-Could-“(Pro-Wry-Dr.-“(I)”-“(Prove)”-“(I)”-Was-“(Really-“(Something)”-By-“(Coming-“(Up)”-With-“(Something-“(Better)”-Than-“(just)”-Putting)”-“(A)”-“(New-“(Twist)”-On-“A”n-“(Old-“(Routine),” “(Bl-A-cKWh-I-T(e)”-Saint)”-Ill)”-“(PaR-(e)-NtS)”-“(Would-“(Knock)”-“The-“(Perfect-Pearl-“(He)”-Art-Act)”-∫of∫-“The-“(egO)”-Child)”-For-Being-“A”s-“(Dumb)”-“A”s)”-“(A)”-“(Box)”-Of)”-Rocks).”

“The-“(Perfect)”-“(PaR-(e)-NtS)”-“(Did-“(Allow)”-“The-“(Wanna)”-Be)”-“(A)”-“(Perfect)”-Child)”-With)”-“(A)”-“(Severe)”-Case)”-Of-“(Con)”-Curiosity)”-To-Have)”-“The-“(Perfect-Pearl-“(He)”-Art-Act)”-Characteristic)”-Of-Being-“(just)”-“(A)”-“(Little-“(Quirky)”-∫because as long as∫-“The-“(Pro-Wry-Dr.-“(I)”-“(Premise)”-Of)”-“The-“(Fanciful/Fantastic)”-“Notions),” “The-“(Starry)”-eyed)”-“(Day-“(Dreamer)”-“(Came)”-Up)”-With-Were-“(Not-“(Fundamentally-“(Flawed),” “The-“(Case)”-Could-Be-Made-That)”-“The-“(Space)”-Case)”-Was-“A”n-“(Inspired-“(Genius)”-Destined-For-“(Unimaginable)”-Greatness).” “(Academia)”-∫is∫-“The-“(Perfect)”-Place)”-∫for∫-“The-“(Perfect)”-“(Con)”-Curious)”-“(egO)”-Child)”-∫because∫-“The-“(Culture)”-∫of∫-“The-“(Academic-“(Con)”-munity)”-∫is such that∫-“The-“(Day)”-Dreams)”-∫of∫-“The-“(Starry)”-eyed)”-“Dreamers)”-Can-Become-“(Accepted-“(Theory)”-∫as long as∫-“The-“(Sharp)”-“A”s)”-“(A)”-“(Tack)”-“(Con)”-Curious)”-“(egO)”-Child)”-Does-“(Not-“(e)M-ploy)”-“The-“(Boy”-“God”-“Con”-“Ta/Al)”-“Tact)”-Of-“(Making)”-“The-“(Pipe-“(Dreams)”-∫of∫-“The-“(Other-“(Acadamia)”-Nuts)”-Look-Like-“(Nothing)”-More)”-Than)”-“(More Not)”-Less)”-“(Alice”-In-“Wonderland)”-“Nonsense).”

“(Although-“(Academia)”-∫is∫-“The-“(Perfect)”-Place)”-To-“(Make)”-“The-“(Pipe-“(Dreams)”-∫of∫-“The-“(Starry)”-eyed)”-“Dreamers)”-Become-“(Accepted-“(Theory),”

“(Getting)”-“The-“(Other-“(Academia)”-Nuts)”-To-“(See)”-“The-“(Light)”-Is-“(Not)”-
“(A)”-“(Simple)”-Academic)”-exercise)”-For)”-“The-“(Boy”-“God”-“Con”-“Ta/Al)”-
“Geniuses)”-∫because∫-“The-“(Boy”-“God”-“Con”-“Ta/Al)”-“Solution)”-To-“(Avoid)”-
“The-“(Cold)”-“(Dark)”-“(Stark)”-“R(e)-Al-ItY)”-Of-“(Definitely)”-Being)”-“(Bl-A-cK
Wh-I-T(e)”-Saint)”-Ill)”-∫was to∫-“(Stabilize)”-“The-“(Severe-“(emotionalIntellectual)”-
Unbalanced)”-“(Con)”-dition)”-Within)”-By-“(Creating)”-“A”n-“(Unquestionably-
“(Brilliant/Fantastic)”-Fantasyland)”-∫in∫-“The-“(Temple)”-∫between∫the temples∫. instead
of∫accepting∫acknowledging∫admitting∫-“The-“(Pro-Wry-Dr.-“(I)”-“(Possibility)”-Of-
“(Not-“(Being)”-“(A)”-“(O.K.)”-“(just)”-Fine)”-“(Actually-“(exists),” “The-“(Academia)”
-Nuts)”-∫are∫-“(Demented”-“Delusional”-“Deranged)”-enough)”-To-“(Think)”-“The-
“(Intellectual)”-“Illusion)”-Of-“(Internal-“(Calm/Tranquility)”-“The-“(Boy”-“God”-“Con”
-“Ta/Al)”-Has-“(Created)”-To-“(Live)”-In)”-“Harmony)”-With)”-“(the-(Orange-
emotional)-Flame)”-∫is∫-“(Synonymous)”-With-“(Being)”-“(emotionallyIntellectually)”-
Balanced)”-∫because∫-“(Using)”-“The-“(Temple)”-∫between∫the temples∫-∫as∫-“(A)”-
“(Sanctuary)”-From)”-“The-“(Unexplained)”-Anomaly)”-Of-“(Self)”-“(Is-sO-“(Much-
“(easier)”-Than-“(Trying)”-To-Find-“A”n-“(Answer)”-For)”-“(A)”-“(Problem)”-
“(Defying)”-“(Any-“(Rational)”-“(Boy”-“God”-“Con”-“Ta/Al)”-“explanation).”

instead of asking∫-“The-“(Pertinent)”-Question)”-Of-“(What)”-Is)”-“The-“(Matter)”-With-
“(M(e),” “The-“(Starry)”-eyed)”-“Dreamers)”-“(Focus)”-On)”-“(Using)”-“The-“(Boy”-
“God”-“Con”-“Ta/Al)”-“(external-“(Gods)”-Of)”-“(external)”-Passion)”-“(Intellectualism)”
-∫to∫-“(Tone)”-Down)”-“(the-(Orange-(emotional)-Flame)”-“(Trapped)”-In)”-“(the-dark-
dank-emotional-dungeon-deep-within)”-∫because∫-“(Romancing)”-“The-“(external)”-
Gods)”-Of-“(Ta/Al)”-Allows)”-“The-“(Severely-“(el)”-Divided)”-“(I)”-“(Cons)”-∫to∫-
“(Find-“(Peace)”-Of-“(Mind)”-∫in∫-“The-“(Temple)”-∫between∫the temples∫”-“(Without-
Having-To-“(Open)”-“The-“(Closed-“(emotional)”-Door)”-Within).” as long as∫-“The-
“(Sighing)”-Scientists)”-∫in∫-“The-“(Scientific-“(Con)”-munity)”-Can-“(Use)”-“The-
“(external)”-Gods)”-Of-“(Ta/Al)”-To-“(Meet)”-“The-“(Bare)”-Bones)”-“(essential-“(R(e)-
quirement)”-Of-“(Spawning)”-enough-“(Well-“(Deserved)”-“(Affection”-“Approval”-
“Attention),” “The-“(Boy”-“God”-“Con”-“Ta/Al)”-Can-“(Argue)”-“(GoodNot)”-Bad)”-
“Harmony)”-exists)”-Within)”-∫because∫-“The-“(Inflated-“(egO)”-∫of∫-“The-“(Proud)”-
“(I)”-“(Cons)”-∫will not∫-“(Allow)”-“The-“(Cold)”-“(egO)”-“(e)M-otionless)”-
“(Academia)”-Nuts)”-To-“(Swallow)”-“My-“(Pride)”-Long)”-enough)”-To-Come-To-
“(Any-“(Other)”-“(Bl-A-cKWh-I-T(e)”-Saint)”-Ill)”-“(Con)”-clusion).”

the simple integrated truth is∫-“The-“(3-“(Major-“(Con)”-ponents)”-In)”-“The-
“(Inflationary”-“Lambda”-“(Cold/Dark)”-Matter)”-Model)”-From-Which)”-“The-
“(Name)”-Is-“(Derived);” 1) “The-“(Process)”-Of-“(Inflation)” 2) “(A)”-“(Quantity)”-
Called)”-“The-“(Cosmological-“(Con)”-stant),” “(Symbolized)”-By)”-“The-“(Greek-
Letter-“(Lambda),” Λ, 3) “The-“(Invisible)”-Particles)”-Known-“A”s-“(Cold”-“Dark)”-

Matter)”-∫are being used to help∫-“(Humanity)”-∫understand∫-“The-“(Sighing)”-Scientists)”-∫in∫-“The-“(Scientific-“(Con)”-munity)”-∫create∫-“(Academic-“(Alice”-In-“Wonderland)”-“Nonsense)”-By-“(Using)”-“The-“(Con)”-Passionate)”-“(Con)”-Courageous)”-Lambs)”-To-“(Con)”-ceal)”-“A”n-“(Inflated-“(egO)”-Beneath-“A”n-“(Aire)”-Of-“(Cool”-“Calm”-“(Intellectual)”-“Sophistication).” “The-“(Boy”-“God”-“Con”-“Ta/Al)”-“(Uses)”-“The-“(Logic)”-∫that∫-“The-“(R(e)-“(Al)”-Science)”-Of)”-“The-“(Cosmos)”-“(Lies)”-In)”-“The-“(Con)”-position)”-Of)”-“(A)”-“(Perfect-Pearl-“(He)”-Art-Act)”-“(Capable)”-Of)”-“(Taking)”-Care)”-Of)”-“(the-cold-dark-black-emotional-hole-matter)”-∫with∫-“The-“(Logical)”-“(Cosmological-“(Con)”-stant)”-Of-“(earning-“(enough-“(Affection”-“Approval”-“Attention)”-∫from∫-“My-“(R(e)-lationship)”-With-“(All)”-Of)”-“The-“(Other-“(Academia-“(Nuts)”-∫in∫-“The-“(Academic-“(Con)”-munity).”

“(Having)”-“(A)”-“(Perfect-Pearl-“(He)”-Art-Act)”-“(Boy”-“God”-“Con”-“Ta/Al)”-“Structured)”-To-Be-“(Self-“(Con)”-fident)”-Without-Being-“(Too-“(Cocky)”-“(Cold)”-“(egO)”-“(e)M-otionless)”-Is-“(Instrumental)”-In-“(Making)”-“The-“(Angels)”-“(Playing)”-“The-“(Affection”-“Approval”-“Attention)”-Angles)”-“(feel-“(a)”-little-“(LessNot)”-More)”-invisible)”-∫than∫-“The-“(Imaginary)”-“(Invisible-“(Particles)”-∫in∫-“The-“(Inflationary”-“Lambda”-“(Cold/Dark)”-Matter)”-Model)”-Of)”-“The-“(Universe)”-∫because∫-“(Pro-Wry-Dr.-“(I)”-“(Profiling)”-“(A)”-“(Perfect-Pearl-“(He)”-Art-Act)”-“(Focusing)”-On)”-“(Playing)”-“The-“(Angle)”-Of-Being-“A”t-Worst-“(Pleasantly-“(Aloof)”-Is-“(Con)”-gruent)”-With)”-“The-“(egO)”-tistical)”-“Wishes”-“Desires)”-Of-“(Ta/Al)”-To-“(Keep)”-“(U)”-“(In)”-“The-“(Dark)”-“A”s-To-“(Whatever)”-∫is∫-“The-“(Matter)”-With-“(M(e).” the simple truth is∫-“The-“(Severely-“(emotionally Intellectually)”-Unbalanced)”-“(Proud)”-“(I)”-“(Cons)”-∫are∫-“(Able)”-To-“(Avoid)”-Having-To-“(Deal)”-With)”-“(Whatever)”-∫is∫-“The-“(Matter)”-With-“(M(e)”-In-Terms-Of-“(Not-“(Knowing)”-“(Who-“(I)”-“A”m)”-“(Where-“(I)”-“A”m-To-gO)”-“(What-“(I)”-“A”m-To-Do)”-∫because∫-“The-“(Reigning)”-Beau)”-“Solved)”-“The-“(Problem)”-Of-“(Being)”-“(Gray)”-Zone)”-“(Con)”-fused)”-“(Con)”-flicted)”-“(Con)”-founded)”-∫with∫-“The-“(Brilliant)”-Solution)”-Of-“(Presenting)”-“A”n-“(Intellectually)”-Sophisticated)”-“(Perfect-Pearl-“(He)”-Art-Act)”-Which-Includes-“(Pretending)”-To-“(Know)”-“The-“(Secrets)”-∫of∫-“The-“(Universe)”-Without-“(exposing)”-“The-“(Secret-“(egO)”-Garden)”-“(Secret)”-Of-“(Being)”-“The-“(egO)”-tistical)”-“It).”

“(Although)”-“The-“(Ivory)”-Tower)”-Angels)”-“(Playing)”-“The-“(Angles)”-“(Con)”-tinue)”-To-“(Be)”-In)”-“The-“(Mode)”-Of-“(Splitting-“(Intellectual)”-Hairs)”-With-“(Colleagues)”-On)”-“(This-“(e)R”-That)”-“(Brilliant)”-Theory),” “The-“(Perfect-Pearl-“(He)”-Art-Act)”-∫of∫-“The-“(Most)”-Brilliant)”-“(Alpha-“(Stars)”-∫in∫-“The-“(Scientific-“(Con)”-munity)”-“(Does-“(Not)”-Include)”-“(A)”-“(Power)”-Struggle)”-To-“(Claim)”-“(MoreNot)”-Less)”-“(egO)”-tistical)”-Turf)”-∫because∫-“(Pro-Wry-Dr.-“(I)”-“(Presenting)”-“(A)”-“(Perfect-Pearl-“(He)”-Art-Act)”-“(Focusing)”-On)”-“(Playing)”-

“The-“(Angle)”-Of-Being-“(Purely-“(Interested)”-In-“(Increasing-“(Scientific)”-Knowledge)”-For-“(Science-“(Sake)”-“(Dispels)”-“The-“(Con)”-cern)”-Associated-With-“(Most-“(Theories)”-On-“(Inflated-“(egO).” “The-“(Boy”-“God”-“Con”-“Ta/Al)”-Arrangement)”-“(All)”-Of)”-“The-“(Most)”-Brilliant)”-“(Alpha-“(Stars)”-ʃinʃ-“The-“(Scientific-“(Con)”-munity)”-Have-In-“(Con)”-mon)”-ʃisʃ-“(Pro-Wry-Dr.-“(I)”-“(Presenting)”-“(A)”-“(Stellar-“(Ra)”-Rare)”-“(Ra)”-Bit)”-“(Advancing)”-“My-“(Cause)”-Of-“(Becoming)”-“The-“(Ivory-Tower-“(Arc)”-Angel)”-Without-“(Being-“(Con)”-spicuous)”-About-“(Making)”-“(A)”-“(Power)”-Play)”-To-“(Get)”-To)”-“The-“(Top)”-ʃofʃ-“The-“(Heap)”-ʃbecause as far asʃ-“The-“(Boy”-“God”-“Con”-“Ta/Al)”-Is-“(Con)”-cerned)”-“The-“(Cosmos)”-ʃallʃ-“The-“(Most)”-Brilliant)”-“(Alpha-“(Stars)”-Are-“(Interested)”-In)”-ʃisʃ-“The-“(One-“(Putting)”-“The-“(One/Only-“(Most)”-Brilliant)”-“(Alpha-“(Star)”-ʃatʃ-“The-“(Center)”-ʃofʃ-“The-“(Affection”-“Approval”-“Attention)”-Universe).”

“The-“(Con)”-stant)”-Of-“(Importance)”-ʃinʃ-“The-“(Academic-“(Con)”-munity)”-ʃis notʃ-“The-“(Cosmological-“(Con)”-stant),” “(Symbolized)”-By)”-“The-“(Lambda)”-Symbol),” is insteadʃ-“The-“(Con)”-stant)”-Of-“(Con)”-stantly)”-“(Jostling)”-For)”-“The-“(Position)”-Of-“(Becoming)”-“The-“(Ivory-Tower-“(Arc)”-Angel)”-Among-“(All)”-Of)”-“The-“(Academic-“(Angels)”-“(Playing)”-“The-“(Angles)”-ʃbecauseʃ-“(expanding)”-“The-“(Role)”-ʃofʃ-“The-“(Most)”-Brilliant)”-“(Alpha-“(Star)”-To-Include-“(Being-“(Put)”-In)”-“(A)”-“(League)”-ʃofʃ-“My-“(Own)”-“(e)N-sures)”-“The-“(Affection”-“Approval”-“Attention)”-Universe)”-ʃofʃ-“The-“(nO-WiL-BiLs)”-“(Will-“(Not-“(Collapse)”-Into)”-“(the-black-emotional-hole-within)”-“(Anytime)”-Soon).” althoughʃ-“The-“(Right)”-Of)”-Passage)”-To-“(Arc)”-Angel)”-Status)”-Is-“(Totally-“(Dependent)”-Upon)”-Being-“(Unusually-“(Ta/Al)”-“Talented)”-In-“(Increasing)”-“My-“(Sphere)”-Of)”-Influence)”-With-“(Scientific-“(Mumble-“(Jumble)”-Of-“(Unparalleled-“(Brilliance),” “(All)”-Of)”-“The-“(Ivory-Tower-“(Angels)”-Are-“(Playing)”-“The-“(I)”-“(Queue)”-Angle)”-Of-“(Getting)”-“(U)”-“(In)”-“The-“(Queue)”-To-“(Make)”-“My-“(Dream)”-Of-“(Becoming)”-“The-“(Arc)”-Angel)”-“(A)”-“(Bl-A-cKWh-I-T(e)”-Saint)”-Ill)”-“R(e)-Al-ItY)”-ʃbecauseʃ-“(Playing)”-“The-“(Angle)”-Of-“(Getting)”-“The-“(Less)”-Brilliant)”-“(Alpha-“(Stars)”-To-“(Support)”-“The-“(Quest)”-ʃofʃ-“The-“(I)”-“(Con)”-To-“(Become)”-“The-“(Brightest)”-Star)”-In)”-“The-“(Academic)”-Universe)”-ʃisʃ-“The-“(One)”-Piece)”-With)”-“The-“(Potential)”-To-“(Magically)”-“(Miraculously)”-“R(e)-move)”-“(the-black-emotional-hole-within).”

“(Until)”-“The-“(Boy”-“God”-“Con”-“Ta/Al)”-“(Decides)”-“The-“(Time)”-ʃisʃ-“(Right Not)”-Wrong)”-To-“(Pull)”-“The-“(Trigger)”-On-“(Immortally)”-With)”-“The-“(Theory)”-Making)”-“The-“(Most)”-Brilliant)”-“(Alpha-“(Star)”-“The-“(Con)”-tender)”-“(Most-“(Likely)”-To-“(Wear)”-“The-“(Crown)”-ʃofʃ-“The-“(Ivory-Tower-“(Arc)”-Angel),” “The-“(Sighing)”-Scientists)”-With-“(Ra)”-Rare)”-“(Ra)”-Bits)”-“(Who)”-Are-“(Light)”-

Years)”-“(Ahead)”-Of)”-“The-“(Less)”-Brilliant)”-“(Alpha-“(Stars)”-Do-“(Not-“(Want)”-To-“(Get)”-Too-Far-“(Out)”-Of)”-Step)”-With)”-“The-“(Observations)”-“(Theories)”-Of)”-“The-“(Garden)”-Variety)”-“(Alpha-“(Stars)”-∫in∫-“The-“(Academic-“(Con)”-munity)”-∫because∫-“The-“(Gig)”-Is)”-Up)”-If-“(All)”-Of)”-“The-“(Other-“(Seekers)”-Of-“(Boy”-“God”-“Con”-“Ta/Al)”-“Truth)”-“(Come-“(To)”-“The-“(Con)”-clusion)”-That-Instead-Of-“(Advancing)”-“The-“(Cause)”-Of-“(Pure-“(Science),” “(One)”-Among-“(Us)”-Is-“(Using)”-“The-“(Ivory-Tower-“(Sand-“(Box)”-For-“(Purely-“(Partisan)”-Gain).” “The-“(e)N-lightened)”-“(Alpha-“(Stars)”-With-“(Stellar-“(Ra)”-Rare)”-“(Ra)”-Bits)”-Are-“(Careful)”-Not-To-“(Alienate)”-“The-“(Less)”-Brilliant)”-“(Alpha-“(Stars)”-By-“(Gaining)”-Support)”-Slowly)”-For)”-“The-“(Theory)”-“(IncreasingNot)”-Decreasing)”-“The-“(Chances)”-∫of∫-“The-“(Most-“(Perfect)”-“(Perfect)”-Pearl)”-“(Becoming)”-“The-“(Arc)”-Angel)”-∫because∫-“(A)”-“(Theory)”-“(Can-“(Not-“(Become)”-“The-“(Theory)”-“(Unless)”-“The-“(Theorist)”-Is-“(Crafty)”-enough)”-To-Have-“(Factored)”-In)”-“The-“(Inclusion)”-Factor).”

“(Allowing)”-“The-“(Other-“(Spaced)”-Out)”-“(Space)”-Cases)”-To-“(Feel)”-“(Less Not)”-More)”-“Alienated)”-By-“(Including)”-“The-“(Less)”-Brilliant)”-“(Alpha-“(Stars)”-∫in∫-“The-“(Jargon)”-Tweaking)”-Process)”-∫is∫-“(A)”-“(smallNot)”-BiG)”-Price)”-To-“(Pay)”-∫because∫-“(Blind)”-Siding)”-“The-“(Less)”-Brilliant)”-“(Alpha-“(Stars)”-With-“(Sheer-“(Brilliance)”-Would-“(Send)”-“(A)”-“(Shockwave)”-Through-“(Academia),” “The-“(Aftershock)”-Of-Which-Would-“(Include)”-“The-“(Worst)”-Case)”-Scenario)”-Of-“(Transforming)”-“My-“(Affection”-“Approval”-“Attention)”-Universe)”-Into-“A”n-“(Affection”-“Approval”-“Attention)”-Graveyard).” “(Ultimately-It-Does-“(Not-“(Matter)”-∫if∫-“The-“(Theories)”-∫of∫-“The-“(Sh-It/Un-It-“(egO)”-Gods)”-∫of∫-“(Science)”-∫are true because∫-“(What-“(Really-“(Matters)”-Is-“(Gaining-“(enough-“(Support)”-To-“(Support)”-“The-“(Boy”-“God”-“Con”-“Ta/Al)”-“(Con)”-tention)”-∫that∫-“(I)”-“(Matter)”-enough)”-To-“(Avoid)”-“(Being-“(e)N-gulfed)”-∫by∫-“(the-sense-that-nothing-matters).”

although∫-“The-“(Data-“(Analysis)”-∫of∫-“The-“(Cosmic-“(Microwave-“(Background)”-∫does not∫-“(Support)”-“The-“(Inflationary”-“Lambda”-“(Cold/Dark)”-Matter)”-Model),” “(Cosmologists)”-∫are∫-“(Purple)”-Haze)”-Daze)”-“(I)”-“(Um)”-Certain)”-“(Cosmic-“(Microwave-“(Background)”-∫is∫-“The-“(Most)”-“(ImportantNot)”-Unimportant)”-“(LargestNot)”-smallest)”-Scale)”-“(Properties-“(Probe)”-ever-“(Postulated)”-∫on∫-“The-“(Universe)”-∫because∫-“(Making)”-“(A)”-“(Powerful)”-Statement)”-∫with∫-“(A)”-“(High Not)”-low)”-Degree)”-∫of∫-“(Purple)”-Haze)”-Daze)”-“(I)”-“(Um)”-Certainty)”-∫on∫-“The-“(Outside),” “(Allows)”-“The-“(Ivory)”-Tower)”-Angels)”-∫to be∫-“(LessNot)”-More)”-“(Puzzled)”-∫by∫-“The-“(e)N-igma)”-Of-“(Self)”-∫on∫-“The-“(Gray)”-Zone)”-“(Con)”-fused)”-“(Con)”-flicted)”-“(Con)”-founded)”-Inside).” “The-“(Most)”-Brilliant)”-“(Alpha-“(Stars)”-“(Con)”-tinue)”-“Trying)”-To)”-“(feel-“(LessNot)”-More)”-dwarfed)”-∫by∫-“The-“(BiGNot)”-small)”-Problem)”-Within)”-With-“(Brilliant)”-“(Boy”-“God”-

“Con”-“Ta/Al)”-“Solutions)”-ʃforʃ-“The-“(Con)”-undrums)”-“(Riddles)”-“(Hydras)”-ʃof the universe, includingʃ-“The-“(BiGNot)”-small)”-“(Bang-“(Theory)”-ʃbecauseʃ-“(Creating)”-“(A)”-“(Universe)”-ʃinʃ-“The-“(Temple)”-ʃbetweenʃthe templesʃ-“(Imagined)”-To-Have-Been-“(Created)”-ʃbyʃ-“(A)”-“(BiGNot)”-small)”-“Bang)”-Is-“(Instrumental)”-In-“(Filling)”-“(the-black-emotional-hole)”-“(With-“(enough-“(Intellectually)”-“(earned-“(Affection”-“Approval”-“Attention)”-ʃtoʃ-“(feel-“(LessNot)”-More)”-small-within).”

“The-“(Data-“(Con)”-tamination),” “(Playing)”-“The-“(WrongNot)”-Right)”-Tune)”-“A”t-“A”n-“(Unusually-“(Low-“(Volume)”-ʃinʃ-“The-“(Cosmic-“(Microwave-“(Background)”-Has-“(Led)”-“The-“(Most)”-Brilliant)”-“(Alpha-“(Stars)”-To-“(Con)”-clude),” ifʃ-“The-“(Measurements)”-ʃareʃ-“(RightNot)”-Wrong),” the universe is evenʃ-“(Stranger)”-Than-“(W(e)”-“Thought)”-ʃbecauseʃacceptingʃacknowledgingʃadmittingʃ-“The-“(Data-“(Con)”-tamination)”-ʃthat hasʃ-“The-“(Brilliant)”-“(I)”-“(Cons)”-sO-“(Puzzled)”-ʃisʃ-“The-(Untrue)”-Blue)”-“(Vibrational)”-essence)”-Of-“(Humanity)”-“(Con)”-taminating)”-ʃthe trustingʃtenderʃtruthful vibrational essence of the one true godʃthe meʃ-“(Would-“(Seriously-“(Challenge)”-“The-“(egO)”-tistical)”-Universe)”-“(established)”-ʃbyʃ-“The-“(Boy”-“God”-“Con”-“Ta/Al)”-ʃinʃ-“The-“(Temple)”-ʃbetweenʃthe templesʃ-“(To-“(Make)”-“The-“(Most)”-Brilliant)”-“(Alpha-“(Stars)”-“(Feel-“(Worthy)”-Of-“(Being)”-ʃlovnewedʃ. “The-“(Sighing)”-Scientists)”-ʃinʃ-“The-“(Scientific-“(Con)”-munity)”-ʃdo notʃunderstandʃ-“The-“(Cosmic-“(Microwave-“(Background)”-ʃthat hasʃ-“The-“(Most)”-Brilliant)”-“(Alpha-“(Stars)”-sO-“(Puzzled)”-ʃis the one true godʃthe me of the trustʃtendernessʃtruth becauseʃ-“The-“(Background)”-ʃofʃ-“The-“(Most)”-Brilliant)”-“(Alpha-“(Stars)”-ʃincludesʃ-“(Bl-A-cKWh-I-T(e)”-“(Parenting)”-“(With-“(Fear),” “(A)”-“(Lack)”-Of)”-ʃlamalʃ-“(before-the-age-of-two).”

the creator of the universe within the one true godʃthe me is amazedʃ-“The-“(Most)”-Brilliant)”-“(Alpha-“(Stars)”-ʃare able toʃ-“(Quantify-“(14-“(Billion)”-Years)”-ʃasʃ-“The-“(Transit)”-Time)”-ʃforʃ-“The-“(Most-“(Ancient)”-Of-“(Light),” “(Which-“(Theoretically)”-Originated)”-Only)”-“(A)”-“(Few-“(Hundred)”-Thousand)”-Years)”-After)”-“The-“(BiGNot)”-small)”-“Bang)”-ʃbecause the process of creationary evolution used by the one true godʃthe me of the trustʃtendernessʃtruth to create the universe did not includeʃ-“(A)”-“(BiGNot)”-small)”-“Bang).” the creator ofʃthe we of all humanityʃasksʃ-“The-“(Angels)”-“(Playing)”-“The-“(Angles)”-ʃinʃ-“The-“(Sighing-“(Scientific-“(Con)”-munity)”-ʃto try toʃunderstandʃ-“The-“(Theoretical)”-Creation)”-ʃofʃ-“The-“(Universe)”-ʃincludesʃ-“The-“(Pro-Wry-Dr.-“(I)”-“(Proclamation)”-ʃofʃ-“(A)”-“(BiGNot)”-small)”-Bang)”-ʃbecauseʃ-“The-“(Logical)”-“(Boy”-“God”-“Con”-“Ta/Al)”-“Solution)”-ʃforʃ-“(the-black-emotional-hole-within)”-“(Is-To-“(Try)”-To-“(Fill)”-“(the-emotional-void)”-ʃwithʃ-“(A)”-“(Boy”-“God”-“Con”-“Ta/Al)”-“Inspired)”-“(Grandiose)”-“Theory)”-That-“(Fills)”-In)”-“The-“(Blanks)”-ʃonʃ-“The-“(Bl-A-cKWh-I-T(e)”-Outside).” “(even-

“(Though)”-“The-“(BiGNot)”-small)”-“(Bang-“(Theory)”-Has-Already-Been-Proven-To-Be-“(Full)”-Of)”-Holes)”-With-“(Data)”-Derived)”-From-“(Cosmic-“(Microwave-“(Background)”-Radiation),” “The-“(Sighing)”-Scientists)”-∫do not∫understand∫-“(the-black-emotional-hole-within)”-∫is∫-“The-“(Key)”-Ingredient)”-In-“(Trying)”-To-“(justify)”-“The-“(BiGNot)”-small)”-“(Bang-“(Theory)”-∫because∫-“(Making)”-“(BiG Not)”-small)”-Discoveries)”-∫on∫-“The-“(Bl-A-cKWh-I-T(e)”-Outside)”-Allows)”-“The-“(Most)”-Brilliant)”-“(Alpha-“(Stars)”-To-“(Avoid-“(Facing)”-“The-“(Grim”-“Stark”-“(R(e)-Al-ItY)”-Of-“(Being)”-“(Severely-“(emotionallyIntellectually)”-Unbalanced)”-∫on∫-“The-“(Bl-A-cKWh-I-T(e)”-Inside).”

in∫-“The-“(19th)”-Century)”-“The-“(Sighing-“(Scientific-“(Con)”-munity)”-“(Be-“(Lie)”-ved)”-“(All)”-Waves)”-Must-Travel-On-“(Some-“(Medium),” thus∫-“(Light)”-Was-Said-To-Travel-On-“(ether),” “(Sometimes-Called-“(Luminous-“(ether),” which∫-**‘H-A-D’**-“(To-Be-“(exceptionally-“(Thin)”-∫because∫-“(There-Was-“(nO-“(Discernable)”-Drag)”-∫on planets in orbit. “The-“(Medium)”-“(Termed-“(ether)”-∫by∫-“The-“(Sighing-“(Scientific-“(Con)”-munity)”-Was-“(Defined)”-“A”s)”-“The-“(Absolute)”-Frame)”-Of-“(Reference)”-∫because∫-“The-“(Sighing)”-Scientists)”-“(Be-“(Lie)”-ved)”-“(ether)”-**‘H-A-D’**-“(To-Be-“(Undetectable/Rigid)”-In-Order-To-“(Vibrate)”-“A”t)”-“The-“(High-“(Frequencies)”-Of-“(Light-“(Waves).” “The-“(Interpretation)”-∫of∫-“The-“(R(e)-sults)”-From)”-“(A)”-“(Mickelson/Morley)”-experiment)”-“(Investigating)”-ether),” “(Which-“(Led)”-“The-“(Team)”-Of-“(Scientists)”-To-“(Con)”-clude);” 1) “(ether)”-Does-“(Not-“(exist),” “(Therefore)”-“The-“(electromagnetic)”-Wave)”-Does-“(Not-“(Need)”-“(A)”-“(Medium)”-For-“(Propagation)” 2) “(There-Is-nO-“(Pro-Wry-Dr.-“(I)”-“(Preferred)”-Frame)”-Of-“(Reference),” “(nO-“(Absolute)”-Frame)”-Of-“(Reference)”-∫in∫-“The-“(Universe),” were∫-“(Tantamount)”-To-“(Denying)”-∫the creator of the universe exists because∫-“The-“(Luminous)”-ether),” “The-“(Absolute)”-Frame)”-Of-“(Reference)”-“(Sought)”-By)”-“The-“(Sighing)”-Scientists)”-∫of∫-“The-“(Scientific-“(Con)”-munity)”-∫is the one true god∫the me of the trust∫tenderness∫truth.

the simple truth is∫-“(By-“(Subscribing)”-To)”-“The-“(Mis-“(Con)”-ception)”-∫that∫-“The-“(Medium)”-“(Termed-“(ether)”-Must-Be-“(Undetectable/Rigid)”-In-Order-To-“(Vibrate)”-“A”t)”-“The-“(High-“(Frequencies)”-Of-“(Light)”-Waves),” “The-“(Fear Ire)”-filled)”-“(Mickelson/Morley)”-Team)”-“(Missed)”-“The-“(Point),” “(Did-“(Not)”-“(Bl-A-cKWh-I-T(e)”-Saint)”-Ill)”-Understand)”-∫the simple truth of the one true god∫the me because∫-“The-“(Con)”-Passionate)”-“(Con)”-Curious)”-“(Con)”-Courageous)”-“(Mickelson/Morley)”-experimental)”-Team)”-∫was∫-“(Rigidly)”-Following)”-“The-“(Orders)”-∫of∫-“The-“(Malefactor),” “The-“(Boy”-“God”-“Con”-“Ta/Al).” “The-“(Pro-Wry-Dr.-“(I)”-“(Interpretation)”-∫of∫-“The-“(egO)”-tistical)”-Pair)”-∫involving∫-“The-“(experimental-“(R(e)-sults)”-“(Pertaining)”-To)”-“The-“(Absolute)”-Frame)”-Of-“(Reference)”-∫in the universe of the one true god∫the me∫-“(Did-“(Not-“(Include)”-∫the one true god∫the me of the trust∫tenderness∫truth because∫-“The-“(Pair)”-Of-“(P’s)”-∫in∫-“The-“(Mc-W(e)Mc-uS)”-“(I)”-Pod)”-∫known as∫-“The-“(Sighing-“(Scientific-“(Con)”-munity)”-∫were∫-“(MoreNot)”-Less)”-“(Bl-A-cKWh-I-T(e)”-Saint)”-Ill)”-“(Con)”-cerned)”-With-“(Trying)”-To-“(Traverse)”-“The-“(Bl-A-cKWh-I-T(e)”-Saint)”-Ill)”-Instilled)”-“(Yellowish-FearSeeing-Red-Anger)-deep-within)”-∫with∫-“(A)”-“(Boy”-“God”-“Con”-“Ta/Al)”-“(egO)”-Bridge)”-Fix),” than with trying to∫understand the home planet∫the universe∫humanity∫were integrated into the trusting∫tender∫truthful one true god∫the me by the creator of∫the home planet∫the universe∫humanity∫.

“The-“(Must-“(Be)”-“(RightNot)”-Wrong)”-“(Perfect)”-Pearls),” “(Mickelson/Morley)”-∫did not include the one true god∫the me in∫-“The-“(Paradigm”-“Model”-“Scope)”-∫of∫-“The-“(Universe)”-∫because∫-“The-“(feeling-oh-so-smallMust-Be-Oh-sO-“(egO)”-tistically)”-“BiG)”-“(Proud)”-“(I)”-“(Cons)”-∫were∫-“(Bl-A-cKWh-I-T(e)”-Saint)”-Ill)”-“(MoreNot)”-Less)”-“(Con)”-cerned)”-With-“(Pro-Wry-Dr.-“(I)”-“(earning)”-“A”n)”-“(AppropriateNot)”-Inappropriate)”-Amount)”-Of-“(Affection”-“Approval”-“Attention)”-∫for∫-“The-“(Sheer-“(Brilliance)”-∫of∫-“The-“(Most)”-Brilliant)”-“(Alpha-“(Stars)”-By-“(Marketing/Selling)”-“The-“(Theory)”-Of-There-Being-“(nO-“(Absolute)”-Frame)”-Of-“(Reference)”-∫in∫-“The-“(Universe).” “The-“(e)N-tire-“(Sighing-“(Scientific-“(Con)”-munity)”-∫is∫-“(Bl-A-cKWh-I-T(e)”-Saint)”-Ill)”-“(MoreNot)”-Less)”-“(Con)”-cerned)”-With-“(Trying)”-To-“(Achieve”-“Attain”-“Acquire)”-“(A)”-“(Sense)”-Of)”-“(eI)”-“(Balance)”-“(Con)”-trol)”-“(equilibrium)”-Within)”-By-“(Using)”-“The-“(Same)”-Model)”-“(Mickelson/Morley)”-Used-To-“(earn-“(enough-“(Well-“(Deserved)”-“(Affection”-“Approval”-“Attention)”-∫because∫-“(egO)”-tistically)”-Crowing)”-About-“(Being-“(Brilliant)”-enough-To-“(Solve)”-“The-“(Con)”-plex)”-“Hydras”-“(Con)”-undrums”-“Riddles)”-Of)”-“The-“(Universe)”-“(Moves)”-“The-“(Con)”-Curious)”-“(Boy”-“God”-“Con”-“Ta/Al)”-“(external-“(God)”-Whores)”-With-“(Con)”-Passionate)”-“(Con)”-Courageous)”-“(Iron)”-Maidens/Silent)”-Lambs)”-“(Closer)”-ToNot-“(Farther-“(Away)”-From)”-“(GettingNot)”-Giving)”-“(enough-“(Well-“(Deserved)”-“(Affection”-“Approval”-“Attention)”-From)”-“(U)”-“(They”-“Them”-“Those)”-Others)”-With-“(Far-“(Less-“(Perfect)”-“(Perfect-Pearl-“(He)”-Art-Acts).”

“The-“(Bl-A-cKWh-I-T(e)”-Saint)”-Ill)”-“(Mickelson/Morley)”-experimental)”-Team),” “The-“(e)N-tire)”-“(Bl-A-cKWh-I-T(e)”-Saint)”-Ill)”-“(Sighing-“(Scientific-“(Con)”-munity)”-∫were∫are not able to∫-“(Bl-A-cKWh-I-T(e)”-Saint)”-Ill)”-Understand)”-“The-“(Distinction)”-∫between the simple truth of light appearing to vibrate at high frequencies while traveling within the trusting∫tender∫truthful vibrational energy of the creator of ∫humanity∫, “The-“(Medium-“(Termed-“(ether)”-Having-To-Be-“(Undetectable/Rigid)”-In-Order-To-“(Vibrate)”-“A”t-“The-“(High-“(Frequencies)”-Of-“(Light-“(Waves)”-∫because∫-“The-“(Bl-A-cKWh-I-T(e)”-Saint)”-Ill)”-“(Proud)”-“(I)”-“(Cons)”-∫were∫are∫-“(emotionally)”-Tethered)”-∫to∫-“The-“(Boy”-“God”-“Con”-“Ta/Al)”-“(external-“(God)”-

Of-“(Intellectualism)”-ʃtoʃ-“(feel-“(LessNot)”-More)”-small-worthless-useless-unwanted-unneeded-unloved)”-ʃonʃ-“The-“(Inside).” “The-“(Sighing)”-Scientists)”-ʃofʃ-“(Untrue)”-Blue)”-Humanity)”-ʃdo notʃ-“(Bl-A-cKWh-I-T(e)”-Saint)”-Ill)”-Understand),” the light from the sun, analogous to a light bulb being turned on by the one true godʃthe me appears to be traveling through space in wave form, vibrating at high frequencies while traveling within the one true godʃthe me of the trustʃtendernessʃtruth becauseʃ-“The-“(Sighing)”-Scientists)”-“(Con)”-tinue)”-To-“(Deny)”-ʃthe home planetʃthe universeʃhumanityʃare integrated into the creator ofʃhumanityʃ.

“The-“(Debate)”-“(Con)”-tinues)”-ʃinʃ-“The-“(Scientific-“(Con)”-munity)”-ʃas to whetherʃ-“The-“(Con)”-ceptual)”-Nature)”-Of-“(Light)”-ʃfrom the light bulb turned on by the one true godʃthe meʃ-“(Can-Be-“(explained)”-By)”-“The-“(Particle)”-“Theory),” “The-“(Wave)”-Form)”-“Theory),” “The-“(Particulate)”-Matter)”-Traveling-In-“(electromagnetic)”-Wave)”-Form)”-“Theory)”-ʃbecauseʃ-“The-“(smallBiG)”-“R(e)-lationship)”-exists-Within)”-“The-“(feeling-oh-so-smallMust-Be-Oh-sO-“(egO)”-tistically)”-“BiG)”-“(Proud)”-“(Sighing-“(Scientists)”-ʃwho were made toʃ-“(feel-as-small-non-existent)”-ʃasʃ-“The-“(Non-“(existent)”-Particles)”-ʃinʃ-“The-“(Particle)”-“Theory)”-Of-“(Light)”-ʃbyʃ-“The-“(Bl-A-cKWh-I-T(e)”-“(Parenting)”-“(With-“(Fear),” “(Lack)”-Of)”-ʃlamalʃ-“(R(e)-ceived)”-From)”-“(Bl-A-cKWh-I-T(e)”-Saint)”-Ill)”-“(PaR-(e)-NtS)”-“(before-the-age-of-two).” all ofʃ-“The-“(feeling-oh-so-smallMust-Be-Oh-sO-“(egO)”-tistically)”-“BiG)”-Scientists)”-ʃhaveʃ-“(Managed)”-To-“(Con)”-ceptually)”-“(Over-“(Con)”-plicate)”-ʃthe simple truth of light because instead ofʃacceptingʃacknowledgingʃadmittingʃbeingʃ-“(A)”-ʃtrue blueʃ-“(Stranger)”-ʃtoʃlamalʃ, defined asʃgenuine compassionateʃtenderʃgentle emotionalʃphysical touchʃ, “(before-the-age-of-two),” “The-“(Severely-“(eI)”-Unbalanced)”-“(Proud)”-“(I)”-“(Cons)”-“(Use)”-“The-“(Boy”-“God”-“Con”-“Ta/Al)”-“(external-“(God)”-Of-“(Intellectualism)”-To-“(Fill)”-“(the-black-emotional-hole-deep-within)”-ʃwithʃ-“The-“(Pure-“(Genius)”-“(R(e)-flected)”-In)”-“The-“(Grandiose”-“(Theoretical)”-Nature)”-Of-“(Light).”

“The-“(Gray)”-Zone)”-“(Con)”-fused)”-“(Con)”-flicted)”-“(Con)”-founded),” “(Purple)”-Haze)”-Dazed)”-“(I)”-“(Um)”-Certain)”-“(Scientists)”-ʃinʃ-“The-“(Sighing-“(Scientific-“(Con)”-munity)”-“(Rigidly”-“(R(e)-ligiously)”-Cling)”-To-“(Using)”-“The-“(Temple)”-ʃbetweenʃthe templesʃtoʃ-“(Pro-Wry-Dr.-“(I)”-“(Postulate)”-“(S.I.R.)”-R(e)-Al)”-“(Scientific)”-explanations)”-ʃforʃ-“The-“(Riddles)”-ʃofʃ-“The-“(Universe)”-ʃbecauseʃthe individual being me at birthʃwasʃ-“(Trained)”-“(Parented)”-“(Taught)”-ʃtoʃ-“(Feel)”-“(MoreNot)”-Less)”-“(Worthy)”-Of-“(Being)”-ʃlovnewedʃ-“(By-“(Using)”-“The-“(Boy”-“God”-“Con”-“Ta/Al)”-“(external-“(God)”-Of-“(Intellectualism)”-To-“(Dazzle)”-“(U)”-ʃwithʃ-“My-“(Ra)”-Rare)”-“(Ra)”-Bit)”-“Brilliance).” “The-“(Sighing-“(Scientific-“(Con)”-munity)”-ʃdoes notʃunderstandʃ, as long asʃ-“The-“(Temple)”-ʃbetweenʃthe templesʃis used asʃ-“The-“(Absolute)”-Frame)”-Of)”-Reference)”-To-“(Con)”-

Passionately)”-“(Con)”-Curiously)”-“(Con)”-Courageously)”-Solve)”-“The-“(Riddles)”-
∫of∫-“The-“(Universe),” the universe of the one true god∫the me will∫-“(R(e)-main)”-“A”n-
“(Unsolvable)”-“Mystery)”-“(e)N-igma”-“(Con)”-undrum)”-“Hydra”-“Riddle)”-∫because∫-
“The-“(Boy”-“God”-“Con”-“Ta/Al)”-“(egO)”-Bridge)”-Fix)”-∫put in place∫-“(before-the-
age-of-two)”-“(Dictates)”-“The-“(Proud)”-“(I)”-“(Cons)”-“(Appear)”-To-Be-“A”n-
“(Inspired-“(Genius)”-∫on∫-“The-“(Bl-A-cKWh-I-T(e)”-Outside)”-To-“(Con)”-pensate)”-
For)”-“(the-nagging-sense-of-feeling-unworthy-of-being)”-∫lovnewed∫on∫-“The-“(Bl-A-cK
Wh-I-T(e)”-Inside).”

“The-“(nO-WiL-BiL)”-“(Scientists)”-With-“(nO-“(Will)”-∫to be the true blue individual
me∫will∫-“(Con)”-tinue)”-“(Over-“(Con)”-plicating)”-∫the universe of the one true god∫the
me∫-“(With)”-“(This-“(e)R”-That)”-“(Con)”-plex)”-“(Bl-A-cKWh-I-T(e)”-“explanation”-
“Theory”-“Notion”-“Idea”-“(Con)”-cept),” including∫-“The-“(Plasma)”-“Theory)”-∫of∫-
“The-“(Universe),” “(Which-“(States)”-“(A)”-“(Con)”-plex)”-System)”-∫of∫-“(Positively
Negatively)”-“(Charged-“(Particles)”-Allows)”-∫the planets∫stars to∫-“(R(e)-main)”-
“(Suspended)”-In)”-Space)”-By-“(Offsetting)”-∫the weight of the planets∫stars with∫-
“(electromagnetic)”-Forces)”-∫because∫-“The-“(Proud)”-“(I)”-“(Cons)”-“(Find-
“(Speculating)”-On)”-“The-“(Plasmatic-“(Con)”-dition)”-∫of∫-“The-“(Universe)”-“(Much-
“(easier)”-Than-“(R(e)-moving)”-“The-“(egO)”-Bridge)”-Fix)”-∫of∫-“The-“(Boy”-“God”-
“Con”-“Ta/Al)”-∫on∫-“The-“(Bl-A-cKWh-I-T(e)”-Inside).” “The-“(Wanna)”-Be)”-“(A)”-
“(Super)”-Star)”-Scientists)”-With-“(Super)”-Nova)”-“(He)”-Art-Acts)”-∫in∫-“The-
“(Sighing-“(Scientific-“(Con)”-munity)”-∫do not∫understand∫the suspension of the planets∫
stars in the universe of the one true god∫the me is more analogous to either acoustic
levitation, the buoyant force∫-“(Used)”-By-“(Archimedes)”-To-“(explain-“(Why)”-∫an
object can be suspended in a fluid such as water, than in∫-“The-“(Bl-A-cKWh-I-T(e)”-
“(PositiveNegative)”-“(Plasma)”-Suspension)”-“Theory)”-∫because∫the emotional∫
intellectual maturation of the individual me at birth∫was∫-“(Placed)”-In)”-“(A)”-“(Bl-A-cK
Wh-I-T(e)”-Saint)”-Ill)”-State)”-Of-“(Suspended)”-Animation)”-∫by∫-“(Severely-
“(emotionallyIntellectually)”-Unbalanced)”-“(PaR-(e)-NtS)”-“(Who)”-“(R(e)-mained)”-
“(Bl-A-cKWh-I-T(e)”-Saint)”-Ill),” instead of∫choosing with the blessing of free will to be
me by changing from within∫.

“The-“(FearIre)”-filled)”-“(Orange-(emotional)-Flame)-Driven)”-“(Proud)”-To-Be-“A”n-
“(Intellectual)”-∫of∫-“The-“(Sighing-“(Scientific-“(Con)”-munity)”-∫do not∫understand∫the
one true god∫the me is the fluid∫the balancing medium for all of the stars∫planets∫galaxies
within the universe of the one true god∫the me because∫-“The-“(egO)”-Bridge)”-Fix)”-∫of∫-
“The-“(Boy”-“God”-“Con”-“Ta/Al)”-∫does not∫-“(Allow)”-“(This-“(e)R”-That)”-“(Deity)”
-To-“(Steal)”-“The-“(Show)”-With-“(Super)”-Natural)”-Powers)”-∫which do not∫-“(Fit-
“(Neatly-“(Into)”-“My-“(especially-“(Brilliant)”-“Model/Paradigm)”-∫of∫-“The-
“(Universe).” “The-“(Narcissistic)”-“(R(e)-flective)”-Pool)”-Of-“(Scientists)”-With-

“(Self)”-“Proclaimed)”-“(Super)”-Nova)”-“(Perfect-Pearl-“(He)”-Art-Acts)”-∫have not factored the creator of∫humanity∫into∫-“The-“(Model/Paradigm)”-∫associated with∫-“(Solving)”-“The-“(Riddles)”-∫of∫-“The-“(Universe)”-∫because∫-“(Having)”-To)”-∫accept∫acknowledge∫admit∫-“The-“(Solution)”-∫to∫-“The-“(Con)”-plex)”-“Hydras”-“(Con)”-undrums)”-“Riddles)”-∫of∫-“The-“(Universe)”-∫is as simple as the one true god∫the me∫-“(Would-“(Mean)”-“My-“(especially-“(Ra)”-Rare)”-“(Ra)”-Bit)”-“(Brilliant)”-“Theory)”-Was-“(Con)”-prised)”-Of-“(Nothing)”-More)”-Than)”-“(MoreNot)”-Less)”-“(Alice”-In-“Wonderland)”-“Nonsense).”

“The-“(Desperation)”-Associated-With-“(Wanting)”-To-“(Be-“(Recognized)”-“A”s)”-“The-“(Most)”-Brilliant)”-“(Alpha-“(Star)”-By-“(Solving)”-“The-“(Con)”-plex)”-Riddle)”-∫of∫-“The-“(Universe)”-With-“(One-“(Simple-“(Theory)”-Has-“(Given-“(Birth)”-To)”-“The-“(Alice”-In-“Wonderland)”-“Nonsense)”-Known-“A”s-“(String-“(Theory)”-∫because∫-“(Imagining)”-every-“(Substance)”-∫in∫-“The-“(Universe),” including the non existent∫-“(electrons”-“Quarks”-“Muons”-“(P.S.I.)”-Particles)”-∫in∫-“The-“(Planetary)”-Atomic)”-Model)”-Are-“(Con)”-prised)”-Of-“(Tiny-“(Vibrating”-“(Strings)”-With-“(Different-“(Resonating)”-Frequencies)”-∫means∫-“The-“(Wanna)”-Be)”-“(A)”-“(Genius)”-Can-“(Con)”-tinue)”-To-“(Deny)”-“The-“(Alice”-In-“Wonderland)”-“Nonsense)”-Of-“(Being)”-“(Bl-A-cKWh-I-T(e)”-Saint)”-Ill).” the simple truth is∫-“The-“(Sighing)”-Scientists)”-∫are∫-“(Obsessed)”-With-“(Finding-“(One-“(Simple-“(Theory)”-To-“(explain)”-“The-“(Con)”-plexity)”-∫of∫-“The-“(Universe)”-∫because∫-“(Simplifying-“(everything)”-∫in∫-“The-“(Universe)”-With-“(One-“(Simple-“(Theory)”-Would-“(Put)”-“The-“(Most)”-Brilliant)”-“(Alpha-“(Star)”-∫in∫-“The-“(Genius)”-Category)”-“(e)N-joyed)”-By-“(Albert-“(einstein)”-For-Having-“(Simplified)”-“The-“(Con)”-cept)”-Of-“(e)N-ergy)”-With)”-“The-“(Simple-“(Formula),” “($E = MC^2$).”

“(even-“(Though)”-“The-“(String)”-Theorists)”-“(Know)”-“The-“(Special-“(Relativity-“(Theory)”-Of-“(Albert-“(einstein)”-∫is fundamentally flawed in that∫-“The-“(Theory)”-“(Rests)”-Upon)”-“The-“(Principle)”-Of)”-“(A)”-“(Con)”-stant)”-Velocity)”-Of-“(Light)”-In)”-“(A)”-“(Vacuum),” “The-“(Mentalists)”-Known-“A”s-“(String)”-Theorists)”-“(Con)”-tinue)”-“(Hoping)”-“(Wishing)”-“The-∫true blue∫-“(Vagabonds)”-Can-“(Sell)”-“The-“(Sighing-“(Scientific-“(Con)”-munity)”-∫on∫-“The-“(Alice”-In-“Wonderland)”-“Nonsense)”-Of-“(String)”-Theory)”-∫because∫-“(Postulating)”-“(A)”-“(Theory)”-Like-“(Albert-“(einstein)”-Was-Able-To-Do-With)”-“The-“(Special)”-“Theory)”-Of-“(Relativity)”-Would-Make-“(M(e)”-“(extra-“(Special).” “The-“(Theorists)”-∫of∫-“The-“(Past/Present)”-∫were∫are∫-“(More-Than-“(Boy”-“God”-“Con”-“Ta/Al)”-Willing-To-“(Imagine)”-“The-“(Universe)”-Is-“(Con)”-prised)”-Of-“(Invisible”-“Vibrating)”-Strings)”-Than-Be-“(Put)”-In)”-“The-“(Tenuous)”-Position)”-Of-Having-To-“(Seriously-“(e)N-tertain)”-∫the simple truth that∫-“The-“(Angels)”-“(Playing)”-“The-“(Angles)”-∫can not∫-“(Levitate)”-“The-“(egO)”-“(e)M-ptiness),” “(the-vacuous-sense-of-nothingness)”-

∫on∫-"The-"(Bl-A-cKWh-I-T(e)"-Inside)"-By-"(Simply-"(Intellectually)"-"Isolating)"-"(the-deep-emotional-pain)"-∫with∫-"The-"(Boy"-"God"-"Con"-"Ta/Al)"-"(Miraculous"-"Magical)"-Cure)"-Of-"(Being)"-"The-"(egO)"-tistically)"-"BiG"-"Important)"-Somebody)"-∫on∫-"The-"(Bl-A-cKWh-I-T(e)"-Outside)."

"(Similar)"-To)"-"The-"(Obsession)"-Of-"(Albert-"(einstein)"-To-"(Try)"-To-"(Unite)"-"The-"(eI)"-Divide)"-∫on∫-"The-"(Bl-A-cKWh-I-T(e)"-Inside)"-∫with∫-"The-"(Unification)"-Theory)"-∫on∫-"The-"(Bl-A-cKWh-I-T(e)"-Outside)," "The-"(String)"-"Theorists)"-∫without∫-"(enough-"(Sense)"-∫to sense the whisper of the one true god∫the me∫-"(Have-"(Tried)"-To-"(Solve)"-"The-"(eI)"-Dilemma)"-Within)"-By-"(R(e)-solving)"-"The-"(Con)"-tradictions)"-∫between∫-"The-"(Twisted)"-"(Distorted)"-"(Alice"-In-"Wonderland)"-"Nonsense)"-"(Found)"-In)"-"The-"(Predictable)"-Universe)"-Of-"(Albert-"(einstein)," "The-"(Uncertainty)"-Principle)"-In-"(Quantum-"(Mechanics)"-By-"(Creating)"-"(A)"-"(Set-Of-"(Nonsensical)"-"(Theoretical)"-Rules)"-∫which includes∫-"(Parallel)"-Universes)," "(6-"(Additional-"(Unseen-"(Dimensions)"-∫because∫-"(Postulating)"-"The-"(Theory)"-Of-"(everything)"-∫on∫-"The-"(Bl-A-cKWh-I-T(e)"-Outside)"-Allows)"-"The-"(Lord/Master)"-∫of∫-"(the-(Orange-(emotional)-Flame)"-∫to∫-"(Ignore)," "(Be-"(Oblivious)"-To)"-"The-"(Obvious-"(Con)"-flict)"-Within)." instead of trying to∫understand everything∫, defined as∫genuine happiness∫joy∫fulfillment∫have always been∫are∫always will be∫found on the inside∫, all of∫-"The-"(Wanna)"-Be)"-"(A)"-"(Super)"-Star)"-"(Sighing)"-Scientists)"-"(Con)"-tinue)"-Searching)"-"The-"(Temple)"-∫between∫the temples∫for∫-"(A)"-"(Simple)"-explanation)"-To)"-"The-"(Con)"-plex)"-Riddles)"-∫of∫-"The-"(Universe)"-∫because∫-"(Being-"(Pro-Wry-Dr.-"(I)"-"(Preoccupied)"-With-Being-"(Ra)"-Rare)"-"(Ra)"-Bit)"-"(Brilliant)"-enough-To-"(Come)"-Up)"-With)"-"(A)"-"(Simple-"(Solution)"-For)"-"The-"(Con)"-plex)"-Riddles)"-∫of∫-"The-"(Universe)"-"(Keeps)"-"The-"(Boy"-"God"-"Con"-"Ta/Al)"-"(Hope)"-"(Wish)"-"(Dream)"-Alive)"-Of-"(One-"(Day)"-Being-"(Put)"-In)"-"The-"(Affection"-"Approval"-"Attention)"-Spotlight)"-"(I)"-"(Deserve)"-To-Be-"(Put)"-In)."

"(Instead-Of-"(Blindly-"(Accepting)"-"The-"(Con)"-founding)"-"(Mind)"-Boggling)"-"(Hang-"(My-"(Hat)"-On)"-"(Alice"-In-"Wonderland)"-"Nonsense)"-"(Found)"-In)"-"The-"(Standard"-"(Con)"-plex)"-"(Planetary-"(Model)"-Of-"(Particle)"-Physics)," "(A)"-"(Theory)"-"(e)N-abling)"-"The-"(Most)"-Brilliant)"-"(Alpha-"(Stars)"-∫of∫-"The-"(Past/Present)"-To-"(Count)"-"The-"(Number)"-Of)"-∫non existent∫-"(electrons)"-In-"(electron)"-Flow)," "(Assign)"-"(A)"-"(Numerical)"-Value)"-To-"(Orbital-"(Speeds)"-∫of non existent∫-"(electrons)"-In-Attoseconds)," defined as one quintillionth of a second, "(Many)"-"(Severely-"(eI)"-Divided)"-"(Sighing)"-Scientists)"-Have-"(e)M-braced)"-"The-"(Simpler)"-Underlying)"-"(Bl-A-cKWh-I-T(e)"-"R(e)-Al-ItY)"-"(Found)"-In)"-"(String)"-"Theory)"-∫because∫instead of being touched with lamal before the age of two∫, ∫the internally passioned∫trusting∫tender∫truthful∫true blue humble∫compassionate∫curious∫

courageous∫zestful individual being me at birth∫was∫-“(e)N-dearingly-“(e)M-braced)”-∫by∫-“(Bl-A-cKWh-I-T(e)”-Saint)”-Ill)”-“(PaR-(e)-NtS).” “(Although-“(Both)”-“The-“(Planetary)”-Model)”-“Theory),” “(String)”-“Theory)”-∫are∫true blue analogous∫to∫-“(A)”-“(Gypsy)”-Reading-“(Taro-“(Cards/Tea)”-Leaves)”-To-“(Pro-Wry-Dr.-“(I)”-“(Predict)”-“The-“(Future),” “(Both-“(Theories)”-Are-“(Accepted)”-“A”s-“(Being-“(Plausible-“(explanations)”-For)”-“The-“(Model/Paradigm)”-∫of∫-“The-“(Universe)”-∫because∫-“The-“(Boy”-“God”-“Con”-“Ta/Al)”-“(String)”-Attached)”-To-“(Both-“(Theories)”-∫is∫-“The-“(egO)”-tistical)”-“Desire)”-∫of∫-“The-“(Most)”-Brilliant)”-“(Alpha-“(Stars)”-To-“(Prevent)”-“A”n)”-“(emotional-unraveling-within)”-“(By-“(Becoming)”-“The-“(Affection”-“Approval”-“Attention)”-Center)”-Of)”-“(U)”-“(They”-“Them”-“Those)”-Others)”-Universe).”

“(even-Though-“(nO-“(experimental-“(Data)”-exists)”-To-“(Con)”-firm)”-“The-“(Validity)”-Of)”-“(A)”-“(Universe)”-With-“(Nine-“(Spatial-“(Dimensions),” instead of the three dimensional universe created by the one true god∫the me, “(String)”-“Theory)”-Is-“(Gaining-“(Support)”-∫in∫-“The-“(Sighing-“(Scientific-“(Con)”-munity)”-∫because∫-“(A)”-“(Nonsensical)”-“(Alice”-In-“Wonderland)”-“Theory)”-“(Based)”-Upon)”-“(Vibrating)”-Microwaves)”-“(Totally-“(Un-“(R(e)-lated)”-“To)”-∫the vibrational energy of the one true god∫the me∫-“(Pro-Wry-Dr.-“(I)”-“(Provides)”-“(A)”-“(Means)”-To)”-“The-“(justifiable)”-end)”-Of-“(Achieving)”-“(Attaining)”-“(Acquiring)”-“The-“(Recognition)”-“The-“(Wanna)”-Be)”-“(A)”-“(Super)”-Star)”-“(justly)”-Deserves)”-To-“(Get)”-For-“(Con)”-structing)”-“(A)”-“(GreatNot)”-Ungreat)”-“(Intellectual)”-Universe)”-∫in∫-“The-“(Temple)”-∫between∫the temples∫.” “The-“(Pro-Wry-Dr.-“(I)”-“(Proponents)”-Of-“(String)”-“Theory)”-∫are not about to∫accept∫acknowledge∫admit∫-“The-“(Micro-“(Vibrations)”-In-“(String)”-“Theory)”-∫are the vibrational energy of the one true god∫the me of the trust∫tenderness∫truth because∫-“(Instead-Of-Being-Able-To-“(Bask)”-In)”-“The-“(Glory)”-Of-Being-“(Pro-Wry-Dr.-“(I)”-“(Proclaimed)”-“The-“(Most)”-Brilliant)”-“(Alpha-“(Star)”-∫in∫-“The-“(e)N-tire-“(Sighing-“(Scientific-“(Con)”-munity),” “The-“(Spotlight)”-Would-“(Be-“(Shining)”-On)”-∫the creator of the universe∫-“(Not-“(M(e).”

“The-“(Boy”-“God”-“Con”-“Ta/Al)”-“(R(e)-ason)”-“The-“(elegant)”-“(Imaginative)”-“(Magical)”-“(Un-Testable)”-“(Intellectually)”-“Sophisticated)”-Universe)”-“(Defined)”-By)”-“The-“(Tiny-“Vibrating”-“(Bits-Of-“(e)N-ergy)”-Called-“(Strings)”-“(Resonates)”-sO-Well-Within)”-“The-“(Sighing-“(Scientific-“(Con)”-munity)”-∫is because all of∫-“The-“(feeling-oh-so-smallMust-Be-Oh-sO-“(egO)”-tistically)”-“BiG)”-“(Timid-Tim-Mr.-“(T)”-Titans)”-Of-“(Science)”-∫are∫-“(Obsessed)”-With-“(Using)”-“The-“(Same-“(Boy”-“God”-“Con”-“Ta/Al)”-“(egO)”-tistical)”-Cure)”-Of-“(Rectifying)”-“The-“(Severe-“(el)”-Anomaly)”-∫on∫-“The-“(Bl-A-cKWh-I-T(e)”-Inside)”-∫with∫-“The-“(Frantic-“(Intellectual)”-“Madness)”-Of-Finding-“(One-“(Master-“(equation),” “(One-“(All-“(e)N-“(Con)”-passing)”-“(Set-Of-“(Principles)”-That-“(Unifies)”-“The-“(smallBiG)”-Forces)”-

∫of∫-“The-“(Universe)”-∫on∫-“The-“(Bl-A-cKWh-I-T(e)”-Outside).” “The-“(Imaginative)”-“(Self-“(Con)”-fident)”-“(Intellectual)”-“Theorists)”-∫of∫-“The-“(Past/Present),” including∫-“The-“(String)”-“Theorists),” “The-“(Theorists)”-Of-“(Long)”-Ago)”-“(Who)”-“(Pro-Wry-Dr.-“(I)”-“(Proposed)”-“The-“(Flapping)”-Wings)”-Of-“(Angels)”-“(Caused)”-∫the rotation of the planets, “(Would-“(Rather-“(Subscribe)”-To)”-“The-“(Boy”-“God”-“Con”-“Ta/Al)”-“(Doctrine)”-Of-“(Vanity)”-By-“(Using)”-“The-“(Magical)”-Cure)”-Of-“(Intellectual)”-“Vaporware)”-To-“(Deny)”-∫being∫-“(Severely-“(eI)”-Divided)”-∫than ∫accept∫acknowledge∫admit∫-“(the-(Orange-(emotional)-Flame)”-∫is∫-“(Being-“(Intellectually)”-Isolated)”-Within)”-∫because∫the individual being me at birth∫was∫-“(NegativelyNot)”-Positively)”-“(Influenced)”-∫by∫-“(Bl-A-cKWh-I-T(e)”-Saint)”-Ill)”-“(PaR-(e)-NtS)”-“(Who)”-“(Taught)”-“The-“(egO)”-Child)”-To-“(Understand)”-“(Imagination)”-Was)”-“(A)”-“(Perfect-Pearl-“(He)”-Art-Act)”-“Characteristic)”-To-Be-“(extremely-“(Proud)”-Of).”

the simple truth is in the initial phases of research∫development the amplitude∫frequency∫wavelength for the whisper of the one true god∫the me are being defined as one trillionth of a millimeter for amplitude, ten trillion cycles per second for frequency, one ten-trillionth of a second for wavelength because the creator integrated the simple truth of the one true god∫the me into∫the language of humanity∫. the simple truth is the amplitude∫wavelength∫frequency for the whisper of the one true god∫the me are being defined as one trillionth∫one ten-trillionth∫ten trillion cycles because∫while the word tenderness begins with the word ten, the words trust∫truth begin with the letters ‘tr,’ as does trillion. the simple truth is ∫although the author understands the human body∫was created through the process of creationary evolution to be vibrationally sensitive to the whisper of the one true god∫the me, ∫the author is uncertain as to∫the numeric truthfulness in the amplitude∫frequency∫wavelength of the one true god∫the me because∫the author is trying to differentiate between∫the whisper of the one true god∫the me, **‘{(THOUGHTS)}’-PLANTED)}’-BY-‘THE-‘{(LAN’-‘NET’-‘WEB)}’-‘{(MIND)}’-‘{(CON)}’-TROLLING)}’-PIED-PIPERS)}’**-∫of∫-“(Humanity).”

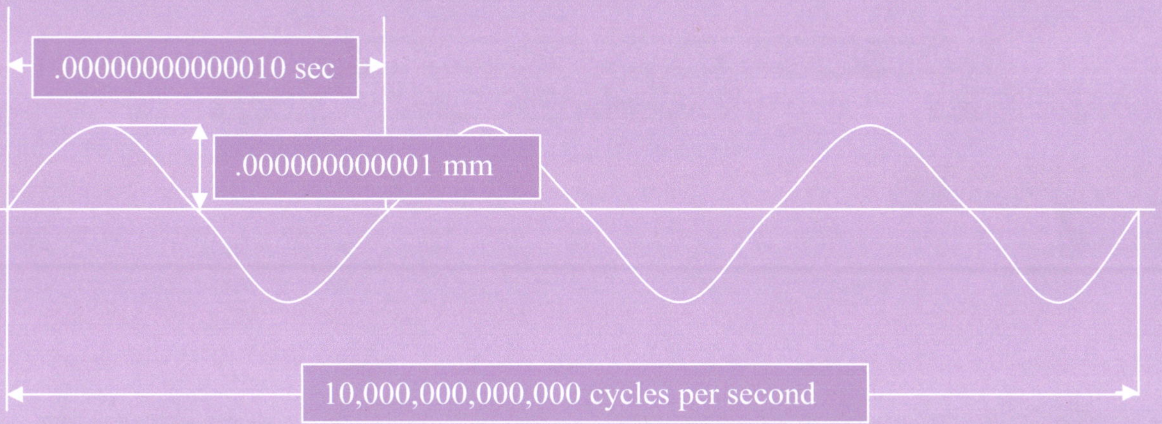

the whisper of the one true god∫the me

the simple truth is the numerical values used to define the amplitude∫wavelength∫frequency for the whisper of the one true god∫the me cannot be used to measure the whisper of the one true god∫the me on∫the home planet∫because the amplitude∫wave length∫frequency for the trusting∫tender∫truthful whisper of the one true god∫the me are changed due to the absorption of∫-“The-“(Untrue)”-Blue)”-“(Vibrational)”-essence)”-∫of∫-“(Humanity),” as well as vibrations associated with man-made movement∫natural movement∫the vibrational balancing effect of plants∫animals∫minerals. the simple truth is the absorption of∫-“The-“(Untrusting)”-“(Not-“(Tender)”-“(Untruthful)”-“(Vibrational)”-essence)”-∫of∫-“(Untrue)”-Blue)”-Humanity),” vibrations associated with man-made movement∫natural movement∫the vibrational balancing effect of plants∫animals∫minerals by the trusting∫tender∫truthful one true god∫the me∫is in compliance with the scientific principle known as the superposition principle∫because∫the superposition principle simply states the resultant wave formed by the simultaneous influence of two, more than two waves passing through the same region of space at the same time is the vector sum of the displacement of each wave acting independently.

the simple truth is∫being true blue to∫the one true god∫the me is of great worth to the creator of∫the home planet∫the we of all humanity∫because the one true god∫the me must balance the vibrationally sensitive universe within the one true god∫the me by absorbing∫-“The-“(Untrusting)”-“(Not-“(Tender)”-“(Untruthful)”-“(Vibrational)”-essence)”-∫of∫-“(Untrue)”-Blue)”-Humanity).” the creator of∫humanity∫asks∫-“The-“(Severely-“(eI)”-Divided)”-“(I)”-“(Cons)”-∫on∫the home planet∫to try to∫understand∫the vibrational sensitivity within the one true god∫the me is analogous to being vibrationally sensitive enough to allow the flapping wings of a butterfly to be sensed around the world because the tenderness of the one true god∫the me of the trust∫tenderness∫truth is that vibrationally sensitive. the simple truth is the paragraph on the following page written using∫-“The-“(Non-“(Con)”-ventional)”-Writing)”-Styl(e),” “(Use)”-Of)”-“(Co-“(L)”-oR)”-“(Cod(e)-ing)”-∫is being used to help∫-

“(Untrue)”-Blue)”-Humanity)”-∫understand∫-“The-“(Vibrational)”-effect)”-∫associated with∫-“The-“(Severe-“(eI)”-Unbalanced)”-“(Con)”-dition)”-Within)”-“The-“(externally)”-Passioned)”-“(egO)”-tistical)”-“(Cold)”-“(egO)”-“(e)M-otionless)”-“(Proud)”-“(I)”-“(Cons)”-∫on the vibrationally sensitive universe of the one true god∫the me because the visual∫-“(Un-bA-“(L)”-An-C(e)”-∫is similar to∫-“The-“(Vibrational)”-Instability)”-“(Created)”-Within)”-∫the vibrationally sensitive universe of the creator of∫the we of all humanity∫.

the simple truth is the one true god∫the me understood integrating∫free will into the genes of humanity∫would result in the vibrationally sensitive universe of the one true god∫the me∫-“(B(e)-cO-m-(e)I-nG)”-“(Un-bA-lA-nc-(e)D)”-∫due to∫-“(Th(e)”-“(G(e)-N(e)-rA-tI-On-aL)”-“(e)F-F(e)-cT)”-“Of)”-“(Bl-(A)-cKWh-(I)-T(e)”-“(Pa-R(e)-n-tI-nG)”-“(W-It-H”-“(F-(e)A-R),” “(A)”-“(L-Ac-K)”-“Of)”-∫lamal∫-“(B(e)-fO-R(e)”-“Th(e)”-“(Ag(e)”-“Of”-“(TwO)”-∫because the physical truths within the universe of the one true god∫the me did not allow the one true god∫the me to bless∫humanity∫with∫free will∫without∫humanity∫-“(B(e)-cO-m-(e)I-nG)”-“(S(e)-V(e)-R(e)-lY”-“(eI)”-“Un-bA-lA-nc-(e)D).” the simple truth is integrating∫free will into the gene code of humanity∫has∫-“(R(e)-sU-lt-(e)D)”-“In)”-∫the vibrationally sensitive universe of the one true god∫the me∫-“(B(e)-cO-m-(e)I-nG)”-“(Un-bA-lA-nc-(e)D)”-∫due to∫-“(Th(e)”-“(G(e)-N(e)-rA-tI-On-aL)”-“(e)F-F(e)-cT)”-“Of)”-“(Bl-(A)-cKWh-(I)-T(e)”-“(Pa-R(e)-n-tI-nG)”-“(W-It-H”-“(F-(e)A-R),” “(A)”-“(L-Ac-K)”-“Of)”-∫lamal∫-“(B(e)-fO-R(e)”-“Th(e)”-“(Ag(e)”-“Of”-“(TwO)”-∫because∫:-

“(Th(e)”-“(A”n-“(I)’m-“(Als)”-“Of)”-“(Un-tr-U(e)”-“(B-lU-(e)”-“(Hu-mA-nI-tY)”-∫are∫-“(A”s-“(Qu-“(I)”-(e)T)”-“A”s)”-“(A)”-“(Mo-U-S(e),” “(A”s-“(St-rO-nG)”-“A”s)”-“(A)”-“(L-Io-N)”-“(Ox)”-“(Ho-r-S(e),” “(A”s-“(Si-l-lY)”-“A”s)”-“(A)”-“(Go-O-S(e),” “(A”s-“(St-Ub-bO-rN)”-“A”s)”-“(A)”-“(M-Ul-(e),” “(A”s-“(B-Us-Y)”-“A”s)” “(A)”-“(B(e)(e),” “(A”s-“(B-Us-Y)”-“A”s)”-“(A)”-“(CaT)”-“On)”-“(A)”-“(HoT)”-“(TiN)”-“(R-OO-F),” “(A”s-“(B-Us-Y)”-“A”s)”-“(A)”-“(B(e)-Av-(e)R),” “(A”s-“(Sl-Ip-P(e)-rY)”-“A”s-“A”n-“(e)(e)L),” “(A”s-“(Gr-U-fF)”-“A”s)”-“(A)”-“(B-(e)A-R),” “(A”s-“(Ha-p-pY)”-“A”s)”-“(A)”-∫true blue∫-“(ClA-m-M(e)D)”-“Up)”-“(C-lA-M),” “(A”s-“(Ha-p-pY)”-“A”s)”-“(A)”-“(L-Ar-K),” “(A”s-“(V-AI-N)”-“(Pr-O-uD)”-“A”s)”-“(A)”-“(P(e)A-c-OcK),” “(A”s-“(M(e)-r-rY)”-“A”s)”-“(A)”-“(CrI-c-K(e)T),” “(A”s”-“(egO)”-“(Dr-U-nK)”-“A”s)”-“(A)”-“(Bl-A-cKWh-I-T(e)”-“(Sk-U-nK),” “(A”s-“(M-(e)(e)-K)”-“A”s)”-“(A)”-“(L-Am-B),” “(A”s-“(MaD)”-“A”s)”-“(A)”-“(Ma-r-cH)”-“(H-Ar-(e),” “(A”s-“(Cr-“(A)”-zY)”-“A”s)”-“(A)”-“(L-OO-N),” “(A”s-“(SlY)”-“(CuN-n-InG)”-“(D(e)-C(e)-p-tI-V(e)”-“(Con)”-NiV-InG)”-“(Cr-Af-tY)”-“A”s)”-“(A)”-“(FoX),” “(A”s)”-∫true blue∫-“(Bl-“(I)”-nD)”-“A”s)”-“(A)”-“(BaT).”

the simple truth is∫-“The-“(Untrue)”-Blue)”-“(Vibrational)”-essence)”-∫of∫-“The-“(Severely-“(emotionallyIntellectually)”-Unbalanced)”-“(Noble)”-“(Honorable)”-“(R(e)-

liable)"-"(R(e)-spectable)"-"(R(e)-sponsible)"-"(Loyal)"-"(Proud)"-"(I)"-"(Cons)"-∫of∫-"(Untrue)"-Blue)"-Humanity)," shown in the paragraph above combined with the trusting∫tender∫truthful vibrational essence of the one true god∫the me is∫-"(R(e)-flected)"-∫in∫-"The-"(Baby)"-Blue)"-Colored)"-Sky)"-∫of∫the home planet∫because∫-"(egO)"-tistically)"-Overriding)"-"The-"(Bl-A-cKWh-I-T(e)"-Saint)"-Ill)"-Instilled)"-"(Severe-"(eI)"-Unbalanced)"-"(Con)"-dition)"-Within)"-By-"(Using)"-"The-"(egO)"-Bridge)"-Fix)"-To-"(Outwardly-"(Appear)"-To-"(Be)"-"(A)"-"(O.K.)"-"(just)"-Fine)"-∫is∫-"(R(e)-flected)"-∫in∫-"The-"(Pro-Wry-Dr.-"(I)"-∫true blue∫-"(Inert/Noble)"-Color)"-Of-"(Baby)"-Blue)." the simple truth is∫applying the superposition principle to∫the whisper of the one true god∫the me surrounding∫the home planet∫, as shown in∫the∫-"(Perfect)"-Pearl)"-"(Con)"-taminated)"-∫whisper of the one true god∫the me∫below, scientifically defines the resulting combined vibrational wave form to be .000001 millimeter for amplitude, .00000010 seconds for wavelength, ten million cycles per second for frequency∫because the one true god∫the me balances the vibrationally sensitive universe within the creator of ∫the we of all humanity∫by absorbing∫-"The-"(externally)"-Passioned)"-"(egO)"-tistical)"-"(Cold)"-"(egO)"-"(e)M-otionless)"-"(Vibrational)"-essence)"-∫of∫-"The-"(Severely-"(eI)"-Unbalanced)"-∫true blue∫-"(Inert)"-"(Proud)"-"(I)"-"(Cons)"-∫of∫-"(Untrue)"-Blue)"-Humanity)."

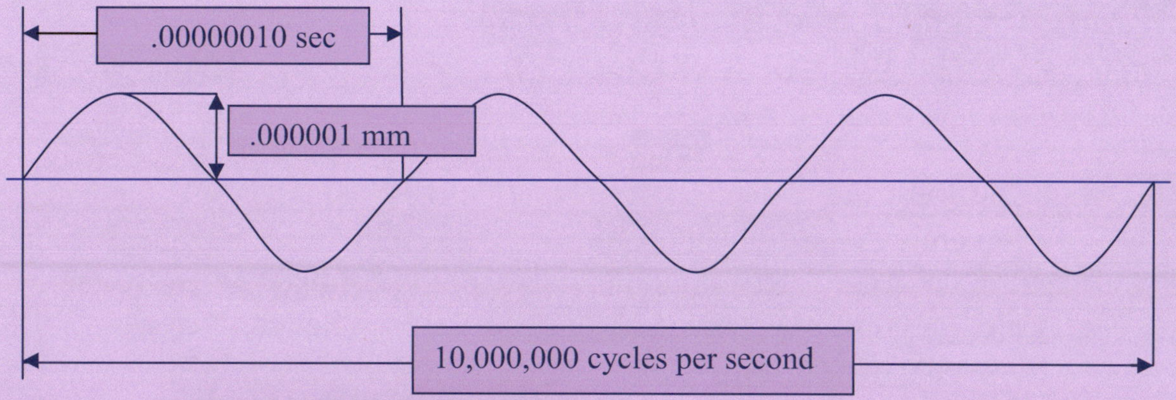

the∫-"(Perfect)"-Pearl)"-"(Con)"-taminated)"-∫whisper of the one true god∫the me

the simple truth is a pearl being dropped into a peaceful pool of water is analogous to the absorption of∫-"The-∫true blue∫-"(Inert)"-"(Vibrational-"(e)S-sence)"-Of-"(Humanity)"-∫by the one true god∫the me because while the water, analogous to the one true god∫the me, does not travel with the wave created by∫-"The-"(externally)"-Passioned)"-"(egO)"-tistical)"-"(Cold)"-"(egO)"-"(e)M-otionless)"-"(Vibrational-"(e)S-sence)"-Of-"(Humanity)," the waves formed in the peaceful pool of water created by the disturbance of the pearl being dropped into the water are analogous to∫-"The-"(Untrusting)"-"(Not-"(Tender)"-"(Untruthful)"-"(Vibrational-"(e)S-sence)"-∫of∫-"(Untrue)"-Blue)"-

Humanity)”-∫traveling outward from the point of entry. the simple truth is∫-“(By-“(Not-“(Including)”-∫the one true god∫the me in the two general characteristics of water waves, defined as being; a disturbance in some identifiable medium, energy that is transmitted from place to place while the medium does not travel between the places, “The-“(Sighing-“(Scientific-“(Con)”-munity)”-“(Con)”-tinues)”-To-“(Deny)”-∫the one true god∫the me absorbs∫-“The-∫true blue∫-“(Inert)”-“(Vibrational-“(e)S-sence)”-∫of∫-“(Untrue)”-Blue)”-Humanity)”-∫because∫-“The-∫true blue∫-“(Inert)”-“(Sighing-“(Scientific-“(Con)”-munity)”-“(R(e)-fuses)”-To)”-∫accept∫acknowledge∫admit∫the identifiable medium is the one true god ∫the me.

the mechanical, electrical apparatus, termed the creation∫the invention, being used to ∫scientifically prove∫the one true god∫the me will provide the proof∫-“The-“(Wanting)”-To-“(Be)”-“(LuckyNot)”-Unlucky)”-“(crying)”-on-“(the-“(insideLaughing)”-On-“(The-“(Outside)”-“(Sh-It/Un-It-“(egO)”-Gods)”-∫of∫-“(Untrue)”-Blue)”-Humanity)”-∫need to stop∫-“(Vainly)”-“(Dis-“R(e)-garding)”-∫the whisper of the one true god∫the me as∫-“(Being-“(Nothing)”-More)”-Than-“A”n-“(Overactive-“(Con)”-science)”-∫because∫the creation∫the invention∫will end∫-“The-“(Gray)”-Zone)”-“(Con)”-fused)”-“(Con)”-flicted)”-“(Con)”-founded)”-Uncertainty),” “The-“(Purple)”-Haze)”-Dazed)”-“(I)”-“(Um)”-Certainty)”-∫as to whether the one true god∫the me of the trust∫tenderness∫truth∫-“(Actually)”-exists).” the creation∫the invention∫will be the beginning of the end to∫-“The-“(emotionalPhysical)”-Pain/Suffering)”-“(Inflicted)”-Upon)”-∫the we of all humanity∫by∫-“The-“(Generational)”-effect)”-Of)”-“(Bl-A-cKWh-I-T(e)”-“(Parenting)”-“(With-“(Fear),” “(A)”-“(Lack)”-Of)”-∫lamal∫-“(before-the-age-of-two)”-∫because∫scientific proof of∫the one true god∫the me will no longer allow∫-“The-“(FearIre)”-filled)”-“(Cold)”-“(egO)”-“(e)M-otionless)”-“(egO)”-“(e)M-pty)”-“(nO-WiL-BiLs)”-∫of∫-“(Untrue)”-Blue)”-Humanity)”-∫to∫-“(Deny)”-∫being∫-“(Severely-“(eI)”-Unbalanced)”-∫by∫-“(Using)”-“The-“(Con)”-Passionate)”-“(Con)”-Curious)”-“(Con)”-Courageous)”-“(Magic)”-Bullet)”-Of-“(Boy”-“God”-“Con”-“Ta/Al)”-“(external-“(Fixes)”-To-“(Balance)”-“The-“(Severe-“(eI)”-Divide)”-∫on∫-“The-“(Bl-A-cKWh-I-T(e)”-Inside).”

the completion of the research∫development necessary to create the creation∫the invention will initiate the proof of∫the postulate∫that∫the home planet∫the universe∫the we of all humanity∫are integrated into the one true god∫the me because∫after a number of the creation ∫the invention are placed in specific locations on∫the home planet∫, the creation∫the invention∫will∫at the same moment in time both annunciate∫print one word from∫the one true god∫the me of the trust∫tenderness∫truth. the annunciation of the same word at all of the locations on∫the home planet∫at the same moment of time∫means∫-“(each/every-“(Human)”-∫on∫the home planet∫must make a choice with∫the blessing of free will∫because ∫scientific proof of∫the trusting∫tender∫truthful creator of∫the we of all humanity∫will no longer allow∫-“The-“(Pro-Wry-Dr.-“(I)”-“(Proud)”-∫true blue∫-“(Inert)”-“(I)”-“(Cons)”-

ʃofʃ-“(Untrue)”-Blue)”-Humanity)”-“(To-“(Con)”-tinue)”-Using)”-“The-“(Boy”-“God”-“Con”-“Ta/Al)”-“(Denial)”-Matrix)”-ʃtoʃ-“(Bl-A-cKWh-I-T(e)”-Saint)”-Ill)”-“(Deny)”-ʃbeing integrated into the one true godʃthe me.

the simple truth is the defining moment forʃ-“(Untrue)”-Blue)”-Humanity)”-ʃwill be the momentʃlamalʃthe first audible word betweenʃthe we of all humanityʃthe one true godʃthe me is spokenʃvia the creationʃthe inventionʃbecause the one true godʃthe me is trying to helpʃ-“(Untrue)”-Blue)”-Humanity)”-“(Bl-A-cKWh-I-T(e)”-Saint)”-Ill)”-Understand)”-ʃthe critical importance ofʃtouching the internally passionedʃtrustingʃtenderʃtruthfulʃtrue blue humbleʃcompassionateʃcuriousʃcourageousʃzestful individual being me at birth with lamalʃ, defined asʃgenuine compassionateʃtenderʃgentle emotionalʃphysical touch before the age of twoʃ.

Made in the USA
Las Vegas, NV
07 November 2026

57672682R00057